Process Analytical Technology and Data Analysis for the Manufacturing of Biologics

Edited by
Amine Kamen, Mario Jardon,
and Rupa Haldavnekar

CRC Press
Taylor & Francis Group
Boca Raton London New York

CRC Press is an imprint of the
Taylor & Francis Group, an **informa** business

Designed cover image: Mario Jardon

First edition published 2027
by CRC Press
2385 NW Executive Center Drive, Suite 320, Boca Raton FL 33431

and by CRC Press
4 Park Square, Milton Park, Abingdon, Oxon, OX14 4RN

CRC Press is an imprint of Taylor & Francis Group, LLC

ISBN: 978-1-032-66291-6 (hbk)
ISBN: 978-1-032-66293-0 (pbk)
ISBN: 978-1-032-66296-1 (ebk)

DOI: 10.1201/9781032662961

Typeset in Times
by KnowledgeWorks Global Ltd.

Contents

Contributors

Mathieu Bourgey
D2R Data Science Director
McGill University
Montreal, Canada

Shawninder Chahal
Department of Chemical and Biological Engineering
University of Ottawa
Ottawa, Ontario, Canada

Laurence Dubeau-Mallette
Department of Bioengineering
McGill University
Montreal, Canada

Julia Puppin Chaves Fulber
Department of Bioengineering
McGill University
Montreal, Canada

Rupa Haldavnekar
School of Electrical and Computer Engineering
University of Oklahoma
Norman, Oklahoma, USA

Javid Jannat
Department of Bioengineering
McGill University
Montreal, Canada

Mario Jardon
Department of Bioengineering
McGill University
Montreal, Canada

Amine Kamen
Department of Bioengineering
McGill University
Montreal, Canada

Malvika Nagrath
Department of Bioengineering
McGill University
Montreal, Canada

Ayyappasamy Sudalaiyadum Perumal
Department of Bioengineering
McGill University
Montreal, Canada

Xingge Xu
Department of Bioengineering
McGill University
Montreal, Canada

About the Editors

Prof. Amine Kamen holds the Canada Research Chair in Bioprocessing of Viral Vaccines at McGill University (retired) and is Researcher Emeritus at the National Research Council of Canada. He is a leading expert in bioprocess engineering, specializing in the development and manufacturing of viral vectors for vaccination and gene and cell therapy. His research has significantly advanced scalable production platforms for adenovirus, lentiviral, adeno-associated virus (AAV), and vesicular stomatitis viral (VSV) vectors, with applications in vaccine, CAR-T cell therapy, and gene editing development. His contributions include the development of serum-free suspension cultures using HEK-293SF and VERO cells, enabling high-yield production of clinical-grade viral vectors. Professor Kamen's work integrates cell culture engineering, process analytical technologies, and quality by design principles to ensure robust and reproducible manufacturing processes. Currently his efforts are directed to advancing mRNA technology for intensification and densification of cost-effective manufacturing processes to enable equitable access. He has authored over 210 peer-reviewed publications and has been instrumental in training the next generation of biomanufacturing professionals through academic and industry collaborations.

Dr. Rupa Haldavnekar is an assistant professor at the University of Oklahoma, School of Electrical and Computer Engineering. She holds a PhD in Biomedical Engineering with focus on surface-enhanced Raman scattering (SERS)–based molecular analysis for cancer diagnosis. Currently Dr. Haldavnekar is developing Raman spectroscopy–based process analytical technology for cancer diagnosis. Her research interests also include the development of orthogonal methods for the investigation of cellular interactions in tumor microenvironment.

Dr. Mario A. Jardon is Faculty Lecturer at the Department of Bioengineering, McGill University. He holds a PhD degree in Chemical and Biological Engineering. He has several years of experience in biopharmaceutical manufacturing. His scientific expertise includes the development and intensification of mammalian cell-based process for recombinant protein production, with a focus on integrating cellular stress responses and metabolism to understand and engineer biological systems responses, with the aim of developing high-performance manufacturing processes used for therapeutic biologics.

Foreword

By Philippe-Alexandre Gilbert

The biopharmaceutical industry stands at a pivotal moment of transformation. Over the past few years, process analytical technology (PAT) and advanced data analytics have moved from theory into practice, fundamentally reshaping how we manufacture biologic medicines. These tools and philosophies—rooted in the regulatory quality by design (QbD) framework—are enabling manufacturers to monitor and control complex processes in real time, rather than relying on end-point testing alone. This shift is not just technical; it carries profound implications for product quality, patient outcomes, and the global supply of biologics.

FROM BATCH TESTING TO REAL-TIME CONTROL

The rising demand for biologics has driven an evolution in biomanufacturing practices toward more consistent quality and greater efficiency. Traditional batch testing, with its inherent delays and blind spots, is giving way to continuous monitoring underpinned by PAT. In practical terms, PAT enables real-time monitoring and control by integrating advanced analytical tools—such as in-line spectroscopy and sensor arrays—and data-driven methodologies directly into upstream and downstream processing. Instead of taking samples for off-line analysis and hoping the batch meets specifications, manufacturers can now watch critical quality attributes in real time and adjust parameters on the fly. This real-time analytics capability shortens production cycles and reduces waste, as processes can be steered to stay within optimal conditions.

Importantly, these approaches improve more than just efficiency—they ensure a higher and more consistent level of quality. Continuous processes augmented with PAT have demonstrated increased efficiency and improved product quality compared to traditional batch methods. Regulators have noted that moving from random end-point sampling to continuous data monitoring can assure quality throughout production. By monitoring critical process parameters and quality attributes continuously, manufacturers detect deviations earlier and prevent substandard product long before a batch is finished.

SMARTER PROCESSES, FASTER RESULTS

The business and patient benefits of these innovations are tangible. Enhanced process understanding—gained by integrating PAT and data analytics into biologics production—directly translates into faster and more reliable manufacturing. For example, implementing PAT in vaccine bioprocessing has led to shorter production times and faster detection of batch failures, while also reducing cost and increasing overall productivity. In general, real-time quality assurance leads to shorter cycle times and fewer stockpiles of inventory, since products can be released as soon as they meet specifications rather than waiting for extensive lab tests. These efficiencies ultimately mean patients get medicines sooner.

Regulators are encouraging this modernization. Adopting technologies like continuous manufacturing—with PAT as a central component—has been shown to reduce quality issues, lower manufacturing costs, and improve the availability of high-quality medicines. In continuous biomanufacturing, an integrated process with fewer hold-points not only speeds up production but also requires a smaller equipment footprint and offers more flexibility to scale production up or down quickly. This flexibility is crucial for reliable supply. The ability to ramp up output or pivot quickly can prevent drug shortages during surges in demand.

The quantitative impact of PAT and analytics is compelling:

- ***Shorter production cycles*:** Continuous monitoring and feedback loops cut batch processing times, accelerating time-to-release by eliminating lengthy off-line tests. Real-time release testing has reduced release times from weeks to days.
- ***Higher yields and fewer failures*:** Automation and PAT lead to more consistent processes, reducing out-of-specification batches and increasing yields. Early detection of deviations allows for corrections that save batches that might otherwise be discarded.
- ***Cost efficiency*:** Optimized resource use, minimized failures, and process control reduce material waste, labor, and energy use, resulting in more sustainable and economical operations.
- ***Rapid scale-up and tech transfer*:** Data-rich models and robust analytics make it easier to replicate or scale processes, enabling faster expansion and localization of manufacturing capacity.

These improvements have moved beyond pilot projects. They are being implemented in commercial facilities and are increasingly viewed as essential to the next generation of biologics manufacturing.

DIGITAL TWINS, AI, AND THE ERA OF "BIOLOGICS 4.0"

A suite of advanced digital technologies is elevating PAT's value even further. Digital twins—virtual models of manufacturing processes—offer powerful capabilities for simulation, prediction, and control. When paired with live plant data, they allow manufacturers to forecast outcomes, test scenarios, and optimize performance without disrupting production.

Digital twins combined with PAT enable a new class of intelligent process control. Machine learning algorithms interpret high-dimensional process data to detect subtle patterns and anomalies. Predictive maintenance algorithms, for example, can anticipate equipment wear before failure occurs. Generative models simulate process changes and predict how they will impact product quality. These insights help operators prevent deviations, minimize downtime, and maintain tight control over complex biologic systems.

This convergence of technologies—sensors, automation, AI, digital twins—is redefining the capabilities of modern biomanufacturing. Processes are not only monitored but increasingly self-optimizing. Data-driven decisions are made in real time. The factory becomes a learning system, growing smarter and more efficient with every batch.

GLOBAL REACH AND INCLUSIVE INNOVATION

The promise of PAT and data-driven manufacturing is not limited to advanced economies. In fact, these technologies have special importance for manufacturers in low- and middle-income countries (LMICs), where building resilient, high-quality production capacity is both a public health imperative and a development opportunity.

The COVID-19 pandemic revealed how fragile global supply chains can be—and how vital it is to have local, responsive manufacturing capabilities. Initiatives aimed at regional vaccine production in LMICs are increasingly incorporating modern process technologies, including PAT and continuous manufacturing. These systems enable efficient, scalable production while ensuring compliance with international quality standards. Digital tools help bridge knowledge gaps, streamline training, and facilitate tech transfer across sites and regions.

Importantly, regulatory guidance is evolving to support these shifts. Global agencies are encouraging adoption of advanced manufacturing systems with flexible pathways for approval and oversight. The message is clear: modern manufacturing approaches, including PAT, are enablers of quality, not barriers. They offer a practical route to both excellence and equity.

A FUTURE OF AGILE, INTELLIGENT MANUFACTURING

The convergence of PAT, real-time data analytics, digital twins, and AI is ushering in an era of agile, intelligent manufacturing. Processes are becoming not just more efficient but more adaptive and resilient. Manufacturing lines that can adjust automatically to maintain quality, scale up rapidly in response to demand, and enable faster product launch are no longer aspirational—they are becoming the norm.

To get there, we must continue investing in collaboration, capability building, and digital literacy. Engineers, data scientists, quality leaders, and regulators must speak a common language of data and control. Cross-disciplinary training, shared standards, and regulatory harmonization will be key. And as innovation accelerates, so too must our commitment to making it accessible to all.

This book makes an essential contribution to that vision. It captures the core principles and latest advancements in PAT and bioprocess analytics, providing practitioners, researchers, and decision-makers with a guide to what's possible—and what's already being done. It speaks to the product, to the process, and to the patient. In doing so, it helps us move closer to a world where every biologic medicine is made with precision, delivered with confidence, and available to all who need it.

Welcome to the era of smart biologics manufacturing. I hope you find the chapters ahead as insightful and inspiring as I did.

Philippe-Alexandre Gilbert, PhD
Biochemical Engineer, Vaccine Development,
Global Health Division, Gates Foundation

1 Introduction
PAT and Digitalization in Pharmaceutical Biomanufacturing

Mario A. Jardon, Amine Kamen, and Rupa Haldavnekar

1.1 THE CURRENT LANDSCAPE

Biopharmaceuticals play a crucial role in modern healthcare, with increasing numbers of biologically derived therapeutics, compared to chemically synthetized small molecules. Today, antibodies, recombinant proteins, vaccines, cell and gene therapies (CGT), and nucleic acid–based therapeutics are not only among the fastest-growing segments of the pharmaceutical market but also among the most transformative segments in terms of patient outcomes. The rise of biologics has redefined what is therapeutically possible—offering cures for previously intractable diseases, extending survival in cancer, providing more target specificity while decreasing side effects, and enabling unprecedented speed in responding to global health emergencies such as the COVID-19 pandemic.

The treatment of numerous instances of cancer, metabolic and immune disorders, chronic conditions, and infectious diseases requires biomanufacturing to bridge the gap between research and actual availability to the public. Biomanufacturing supports the development and large-scale manufacturing of complex biologic therapies such as antibodies (monoclonal, bispecific, antibody–drug conjugates, and fragment-based therapeutics), CGT, and messenger RNA (mRNA)-based technologies.

The term "Industry 4.0" is widely used to refer to the deep changes that digitalization and connectivity are having on manufacturing, in close association with artificial intelligence (AI), data analytics, the internet of things (IoT), and cyber-physical systems (CPS). In this context, Industry 3.0 normally refers to automation enabled by electronics and computer technology, Industry 2.0 to mass production enabled by the implementation of assembly lines and electricity, and Industry 1.0 to the changes produced by the mechanization of production made possible by the steam engine.

Biopharmaceutical manufacturing is not isolated from the current trends driving innovation in other sectors, but the implementation of enabling technologies in this field involves major challenges due to the inherent variability of living systems. For therapeutic biologics to reach successful commercialization, it is paramount that they are manufactured safely, consistently, and at scale. As numerous biotherapeutics have successfully transitioned from research and development phases into full-scale

DOI: 10.1201/9781032662961-1

commercialization, regulatory agencies and industry have increasingly embraced the application of Quality by Design (QbD), a systematic approach to development that begins with predefined objectives and emphasizes product and process understanding.[1–4] From the QbD perspective, the quality of a product is the result of a well-defined and controlled manufacturing process, designed around a thorough understanding of the factors that impact product performance. The grounding of this approach on scientific and risk management principles provides the basis for the implementation of efficient process control and continuous improvement strategies. The overall workflow of QbD can thus be systematized as a cycle of monitoring, understanding, control, and improvement, applied to products and processes, as summarized in Figure 1.1. A more thorough description of the methodologies and specific deliverables of QbD will be developed in the next chapter.

In parallel, Process Analytical Technology (PAT) tools are employed to monitor and control manufacturing in real time, supporting the implementation of QbD by enabling continuous quality assurance and facilitating robust, flexible, and efficient production platforms.[5,6] Biomanufacturing is not merely the industrial realization of a therapeutic concept; it is an intricate convergence of biology, chemistry, engineering, materials science, and regulatory science. Unlike small molecules, where chemical synthesis routes are typically well defined and reproducible, the production of biologics is highly sensitive to cell line characteristics, culture conditions, raw material variability, and downstream purification processes. The relatively accepted expression "the product is the process" underscores that quality is not something tested into biologics at the end of production—it must be built into every step of the manufacturing process.

It is within this context that **PAT and data-driven manufacturing** have emerged as transformative paradigms. They represent both a response to the increasing complexity of biologics and a proactive strategy for ensuring efficiency, reliability, and regulatory compliance in biomanufacturing. This book seeks to explore these paradigms in depth. Before doing so, it is necessary to situate them within the broader landscape of pharmaceutical biomanufacturing—its opportunities, challenges, and evolving demands.

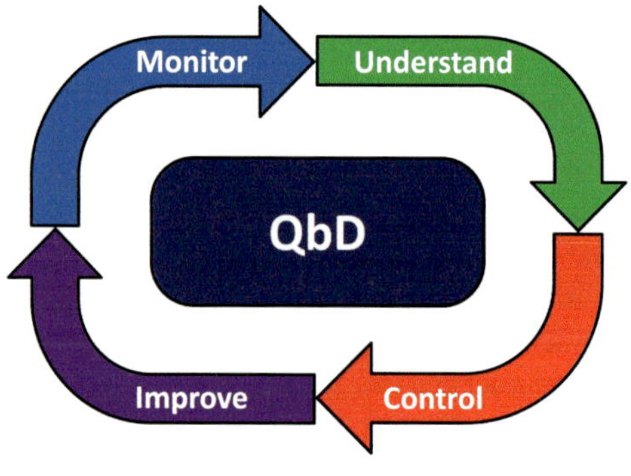

FIGURE 1.1 QbD workflow.

Although the concept of PAT is over 20 years old, there are numerous tools and software platforms dedicated to PAT implementation in pharmaceutical manufacturing; most current applications remain limited to small molecule drugs, particularly in simpler process steps such as drying and tableting, where only a few parameters require monitoring. As per the Food and Drug Administration (FDA) definition, PAT is "a system for designing, analyzing, and controlling manufacturing through timely measurements (i.e., during processing) of critical quality and performance attributes of raw and in-process materials and processes, with the goal of ensuring final product quality".[7] Since the term "analytical" is viewed broadly, encompassing any chemical, physical, microbiological, mathematical, or risk analysis, the goal of implementing PAT serves as a tool to understand and control the manufacturing process rather than testing the quality of the final product.

1.2 THE BIOLOGICS MARKET: SCALE AND GROWTH

Biomanufacturing is part of a vast and rapidly expanding global market. By the end of 2025, the global biologics market was estimated at USD 450 billion and was projected to more than double, reaching USD 1.08 trillion by 2035, growing at a compound annual rate of approximately 9%.[8] Other industry forecasts are aligned—with one citing a base value of USD 487 billion in 2025, rising to USD 1.14 trillion by 2034, at a compound annual growth rate (CAGR) near 10%.[8] Within this broader market, the biologics manufacturing segment (covering cell-based production, contract manufacturing, continuous platforms, and single-use systems) was valued at around USD 198 billion in 2024 and is expected to reach USD 519 billion by 2034, growing at a ~10 % CAGR.[9] Meanwhile, next-generation biomanufacturing—the modular and continuous systems powering advanced biologics—represents a fast-growing sub-sector, valued at approximately USD 26.6 billion in 2024, and projected to rise to USD 56.4 billion by 2032.[10] North America currently leads in absolute scale, accounting for a substantial portion of revenue, but Asia-Pacific—especially China, India, and South Korea—is the fastest-growing region, often posting double-digit CAGR values across all these markets.[10] These figures illustrate both the economic magnitude and the growth trajectory of biologics and biomanufacturing—and set the stage for why PAT and data-driven manufacturing are no longer discretionary but essential.

1.3 BIOMANUFACTURING TODAY: FOUNDATIONS AND EVOLUTION

Modern biomanufacturing builds upon decades of advances in molecular biology, cell culture engineering, and industrial-scale pharmaceutical production. Mammalian expression systems, particularly Chinese hamster ovary (CHO) cells, remain the workhorse of therapeutic protein production due to their ability to perform human-like post-translational modifications. Microbial systems such as *Escherichia coli* and yeast continue to serve important roles in producing smaller proteins, enzymes, and vaccines. Insect cell systems and plant-based platforms have also expanded the toolbox of biologics production. RNA-based therapeutics are also regulated as biologics, since living systems are used in their manufacturing process, namely bacteria to produce template DNA required for *in vitro* transcription (IVT) operations. Whereas numerous efforts are shifting RNA products toward full synthetic production excluding living systems, the PAT and QbD principles described in this volume can be applicable to their manufacturing due to their complex nature.

Facilities have historically relied on large, stainless-steel bioreactors with batch or fed-batch processes. These systems are capital-intensive, slow to adapt, difficult to modify, and dependent on extensive manual oversight. However, the past two decades have seen a steady transition toward **single-use bioreactors, modular facilities, and continuous processing approaches**. These innovations reflect an industry seeking flexibility, speed, and cost efficiency—qualities that are essential in a world of rapidly evolving therapeutic needs.

The regulatory landscape has also evolved in parallel. The U.S. FDA and the European Medicines Agency (EMA) have championed **QbD** and **PAT frameworks**, encouraging manufacturers to move away from empirical, end-product testing toward science- and risk-based approaches. This shift represents not only a change in manufacturing philosophy but also a call for a new generation of analytical and data tools capable of providing real-time, actionable insights into complex bioprocesses.

1.4 OPPORTUNITIES DRIVING BIOMANUFACTURING INNOVATION

The promise of biomanufacturing extends well beyond current achievements. Several current trends represent powerful opportunities that are shaping its future trajectory:

1. ***Personalized Medicine and Small-Batch Manufacturing***: Advances in genomics, immunology, and bioinformatics are ushering in an era of individualized therapies. CAR-T cells, patient-specific vaccines, and precision protein therapeutics require manufacturing systems that are agile, modular, and capable of rapid turnaround. This shift makes industry lean toward small-batch, high-diversity production models supported by PAT-driven quality assurance.
2. ***Continuous Manufacturing***: Continuous bioprocessing holds the potential to reduce costs, accelerate production cycles, and improve quality consistency. Unlike traditional batch operations, continuous systems integrate upstream and downstream processes, supported by real-time analytics and control. PAT is indispensable for realizing this vision, enabling monitoring of critical quality attributes (CQAs) and critical process parameters (CPPs) in real time.
3. ***Digitalization and Data Integration***: Biomanufacturing is not an exception in its evolution toward what is called Industry 4.0, through the increased integration of physical equipment with digital technologies. This trend is observed in the increasing adoption of **digital twins**, AI-driven predictive models, and integrated data environments. These tools not only optimize process parameters but also enable proactive deviation management and predictive maintenance. The fusion of PAT and advanced analytics is at the heart of this digital transformation.

4. ***Global Expansion and Equitable Access***: Emerging economies such as India, China, and South Korea are rapidly building biomanufacturing capacity, often focusing on biosimilars and vaccines. Similar trends are observed in low- and medium-income countries (LMICs), including South Africa, Senegal, and Indonesia, and are supported through programs funded by major charity organizations such as the Gates Foundation and managed by WHO and PATH organizations. This democratization of manufacturing presents opportunities for cost reduction and broader therapeutic access, provided that consistent quality and regulatory harmonization are maintained. The emergence of biosimilars also has an impact on the availability of established first-in-class biotherapeutics.
5. ***Pandemic Preparedness***: The COVID-19 pandemic demonstrated the need for rapid-response platforms such as mRNA vaccines. Investing in flexible manufacturing systems, supported by analytical and data infrastructures, is now a global priority for pandemic preparedness and biothreat resilience.

1.5 CHALLENGES CONFRONTING THE FIELD

Despite these opportunities, biomanufacturing faces formidable challenges that underscore the importance of PAT and data-centric approaches. From a biomanufacturing perspective, the main challenge lies in the analysis of processes where complex, dynamic biological systems govern product quality, thereby making real-time monitoring and control significantly more difficult.[1] There are thousands of interconnected biochemical reactions that occur simultaneously in a bioreactor, as living cells consume nutrients generating a wide range of metabolites.[11] Bioreactors are closed, controlled environments designed to cultivate microorganisms or cells under precisely defined conditions that can be optimized for cell viability, process reproducibility, and product-specific productivity. To steer the intricate biochemical reaction networks within these biological systems toward the desired outcomes, it is essential to understand and regulate the complex interactions between cellular activity, nutrient availability, and environmental conditions. Effective control of these variables is key to ensuring consistent and efficient bioprocess performance. The CQAs of the final product are significantly influenced by the complexity of biomanufacturing processes. Moreover, the therapeutic itself is typically a large biomolecule accumulated in the cells or secreted in the surrounding extracellular environment—a complex mixture of components, including residual elements from the original culture medium and various host cell proteins. The presence of these diverse and fluctuating substances makes real-time, on-line measurement of both CQAs and many CPPs highly challenging, posing a significant barrier to effective process monitoring and control in biomanufacturing. Some of these challenges can be summarized as follows:

1. ***Process Complexity and Variability***: Living systems are inherently variable, and small perturbations in raw materials, culture conditions, or purification steps can significantly alter product quality. Traditional quality control methods, reliant on end-product testing, are insufficient to guarantee consistency.

2. ***Scalability and Cost***: While laboratory-scale production is relatively straightforward, scaling to commercial levels often introduces unforeseen issues. Autologous CGTs, in particular, face high costs and scalability bottlenecks.
3. ***Regulatory Uncertainty***: Though agencies support QbD and PAT, global regulatory harmonization remains incomplete. Manufacturers often navigate a patchwork of requirements that can slow innovation.
4. ***Workforce and Skills Gap***: Biomanufacturing requires interdisciplinary expertise in biology, engineering, and data science. The current workforce often lacks sufficient training in integrating analytics and digital technologies into bioprocesses.
5. ***Sustainability and Supply Chain Vulnerabilities***: The increasing reliance on single-use systems raises environmental concerns, while supply chains for specialized raw materials (e.g., cell culture media, lipids, viral vectors) remain fragile and geographically concentrated.

1.6 THE ROLE OF PAT AND DATA ANALYSIS

The variety of PAT tools used in biomanufacturing includes in-line, at-line, and offline PAT. In-line sensors deliver continuous, real-time data directly from bioprocess without the need for intermittent sampling. These sensors function either by making direct contact with the process medium—referred to as invasive sensing—or by remaining physically separated from the medium through a transparent barrier, such as a glass or quartz window, which allows for non-invasive monitoring. When configured in a bypass arrangement, where the sensor is positioned along a diverted stream of the process flow, they are often termed on-line sensors. Regardless of the configuration, the key advantage of in-line and on-line sensors is their ability to provide uninterrupted streams of data, which are essential for enabling real-time, automated control strategies. This facilitates immediate responses to process deviations and supports the implementation of advanced control techniques such as feedback and feedforward loops, thereby enhancing overall process stability and product quality. In contrast, at-line sensors operate in close proximity to the process vessel but require discrete sampling events at predetermined time intervals. These samples may be collected manually by an operator or through automated sampling systems and are then analyzed using dedicated instrumentation near the production area. While at-line measurements are generally faster than traditional laboratory analyses, they involve time delays between sampling and data acquisition inherent to the analytical process. As a result, at-line data are primarily suited for process monitoring and trending rather than real-time control, especially in fast-changing bioprocess environments. Off-line measurements represent the most traditional form of process analysis, wherein samples are manually or automatically collected from the process vessel and transported to a centralized laboratory for testing. These analyses often employ highly sensitive and specific techniques, such as chromatography or mass spectrometry, that require longer sample preparation and processing times. Consequently, the significant delay between sample collection and result generation limits the utility of off-line data for dynamic process control. Instead, off-line measurements are typically used for batch release testing, product characterization, or validation purposes rather than for ongoing, real-time control and decision-making within the manufacturing process.[12]

Against this backdrop, and especially in its in-line or on-line versions, **PAT emerges as both a necessity and an opportunity**. By embedding analytical tools—spectroscopy, chromatography, mass spectrometry, advanced sensors—directly into manufacturing workflows, PAT enables real-time monitoring and control of processes. Coupled with advanced data analysis, including statistical design of experiments, multivariate statistical modeling, and machine learning, PAT presents new opportunities for dynamic process optimization, rapid troubleshooting, and, ultimately, the goal of **real-time release testing**.

Data analysis further extends the value of PAT by enabling

- Predictive modeling of the complex associations between process conditions and physiological responses of the living systems being manipulated, and how that affects performance and quality attributes.
- Early detection of process deviations that may negatively affect product performance and therefore means to proactively correct or even prevent undesirable effects.
- Integration of multi-sensor and even different kinds of omics data into unified process control systems, increasing performance and efficiency.
- Creation of digital twins that can simulate, test, and refine bioprocesses before physical implementation, and interaction with the physical system once implemented.

Together, PAT and data analytics redefine biomanufacturing not as a rigid, empirical exercise, but as an adaptive, knowledge-driven system.

1.7 PURPOSE OF THIS BOOK

The remarkable growth of biologics and the rapid expansion of biomanufacturing capacity highlight both an opportunity and a gap. While the industry is investing billions of dollars into next-generation platforms, there remains a pressing need for skilled professionals who can integrate biological knowledge with engineering principles, analytics, and data science.

Currently, few resources provide a unified introduction to PAT and data-driven biomanufacturing that is accessible to both academic learners and industry practitioners. Undergraduate and master's students often encounter fragmented discussions of PAT—embedded within courses on biotechnology, pharmaceutical engineering, or regulatory science—without a clear narrative that links theory, tools, and application. Likewise, engineers and scientists in industry who seek to transition into biomanufacturing often face a steep learning curve, particularly when adapting from traditional pharmaceutical or chemical manufacturing to biologics, where variability, complexity, and regulatory expectations are paramount.

This textbook has been written to address that gap. Its goals are threefold:

1. *For Undergraduate and Master's Students*: To provide a structured foundation in PAT principles, analytical tools, and data analysis methods, facilitating readiness for careers in biopharma manufacturing and regulatory science.
2. *For Industry Professionals in Transition*: To serve as a practical guide that bridges prior expertise in engineering, analytics, or chemistry with the specific challenges of biologics manufacturing.
3. *For the Broader Biomanufacturing Ecosystem*: To offer a reference that integrates technical, regulatory, and strategic perspectives on how PAT can transform manufacturing efficiency, reliability, and sustainability.

By framing PAT and data analysis not as add-ons, but as integral to the future of biomanufacturing, this book seeks to prepare the next generation of professionals—whether students entering the field or experienced practitioners evolving with it.

In this context, the chapters that follow the present chapter are organized to provide a logical and comprehensive pathway through the landscape of PAT and data analytics for biologics manufacturing. We begin with the QbD framework, which establishes the scientific and regulatory foundation for modern bioprocess development in Chapter 2. Building on this, we explore PAT tools for process monitoring in Chapter 3, followed by an essential analytics framework for data analysis, highlighting how raw process data can be transformed into actionable insights (Chapter 4). The text then examines the application of machine learning in PAT (Chapter 5), before moving into process control and process modeling, where the integration of analytics with decision-making is emphasized (Chapters 6 and 7). Finally, we turn to data management, storage, and security, reflecting the realities of digital transformation in biomanufacturing (Chapter 8) and conclude with future perspectives on PAT for biologics, pointing toward the opportunities and challenges that lie ahead (Chapter 9). Additionally, a case study on mRNA manufacturing, to anchor the concepts of CQAs of a specific drug product and PAT for monitoring and control of the CPPs is discussed in Chapter 10. Together, these chapters are designed to provide students, educators, and industry professionals with both a conceptual foundation and a practical toolkit for understanding and applying PAT in the rapidly evolving field of biologics manufacturing. Ultimately, the purpose of this book is to equip readers with the knowledge and mindset needed to thrive in an era where data, analytics, and adaptive manufacturing define success in biologics production.

REFERENCES

1. Daniel, S., Kis, Z., Kontoravdi, C. & Shah, N. Quality by design for enabling RNA platform production processes. *Trends Biotechnol* **40**, 1213–1228 (2022). https://doi.org/10.1016/j.tibtech.2022.03.012
2. Lionberger, R. A., Lee, S. L., Lee, L., Raw, A. & Yu, L. X. Quality by design: concepts for ANDAs. *AAPS J* **10**, 268–276 (2008). https://doi.org/10.1208/s12248-008-9026-7
3. Yu, L. X. Pharmaceutical quality by design: product and process development, understanding, and control. *Pharm Res* **25**, 781–791 (2008). https://doi.org/10.1007/s11095-007-9511-1

4. Yu, L. X. *et al*. Understanding pharmaceutical quality by design. *AAPS J* **16**, 771–783 (2014). https://doi.org/10.1208/s12248-014-9598-3
5. Bakeev, K. A. *Process analytical technology: spectroscopic tools and implementation strategies for the chemical and pharmaceutical industries*. 2nd edn (Wiley, 2010).
6. Kim, E. J., Kim, J. H., Kim, M. S., Jeong, S. H. & Choi, D. H. Process analytical technology tools for monitoring pharmaceutical unit operations: a control strategy for continuous process verification. *Pharmaceutics* **13** (2021). https://doi.org/10.3390/pharmaceutics13060919
7. FDA. (https://www.fda.gov/regulatory-information/search-fda-guidance-documents/pat-framework-innovative-pharmaceutical-development-manufacturing-and-quality-assurance, 2004).
8. Precedence Research. (https://www.precedenceresearch.com/biologics-market?, 2025).
9. Analytics. (https://www.usdanalytics.com/industry-reports/biologics-manufacturing-market?, 2025).
10. Bridge, D. (ed Sachin Pawar) (https://www.databridgemarketresearch.com/reports/global-next-generation-biomanufacturing-market?utm, 2025).
11. Streefland, M., Martens, D. E., Beuvery, E. C., Wijffels, R. H. Process analytical technology (PAT) tools for the cultivation step in biopharmaceutical production. *Eng Life Sci* 13(3), 212–223 (2013). https://doi.org/10.1002/elsc.201200025
12. Lourenço, N. D., Lopes, J. A., Almeida, C. F., Sarraguça, M. C. & Pinheiro, H. M. Bioreactor monitoring with spectroscopy and chemometrics: a review. *Anal Bioanal Chem* 404, 1211–1237 (2012). https://doi.org/10.1007/s00216-012-6073-9

2 The QbD Approach for Manufacturing of Biologics

Ayyappasamy Sudalaiyadum Perumal and Mario A. Jardon

2.1 THE REGULATORY PROCESS AS A FRAMEWORK FOR QbD

On its way from discovery to market, a biopharmaceutical product requires a series of approvals by the respective health authorities. Figure 2.1 summarizes this regulatory process and how the various deliverables and methodologies used by the Quality by Design (QbD) approach can support a product's regulatory compliance. While not being a regulatory requirement, it is strongly encouraged by regulatory bodies.

Before a product can be tested in human subjects, preclinical research aims at demonstrating its safety using cell and animal models. This testing allows for the identification of potential toxicity and provides an initial assessment of its efficacy and mode of action. Once the respective Investigational New Drug (IND) application is authorized, clinical trials (human testing) can proceed in its various phases. Phase I clinical trials are conducted in a relatively small population (typically fewer than 100 participants) to assess the drug's safety and tolerability, as well as its pharmacokinetics and pharmacodynamics (PK/PD). The main objective of Phase II trials

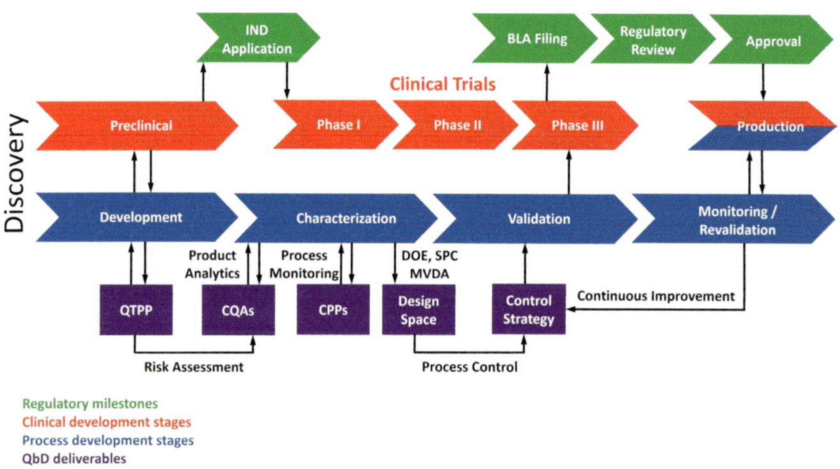

FIGURE 2.1 Regulatory roadmap for biopharmaceuticals and integration of QbD in the process development (PD) stages.

DOI: 10.1201/9781032662961-2

is to evaluate the immunogenicity, efficacy, effective dose, and route of administration, and they are performed on a larger population (in the order of a few hundred). When a product reaches Phase III of clinical testing, the pool of participants can be expanded to a few thousand, and the main objective shifts to definitively assessing the safety and efficacy of the product. Application for market approval is then done through a Biologics License Application (BLA), which undergoes regulatory review and approval if evidence is considered conclusive.

2.2 THE DELIVERABLES OF QbD

2.2.1 Quality Target Product Profile (QTPP)

The background PD activities that support clinical testing include development, characterization, and validation. Normally, PD for preclinical testing involves determining the methods for production at a larger scale, as well as the design of the delivery system or formulation in which the final product will be administered. Analytical methods are required from a very early stage to assess the key features of product performance. It is at this early stage of PD that a prospective summary of the quality features of a product is ideally established. This summary is called the Quality Target Product Profile (QTPP), and it serves as a strategic basis to align PD activities [1]. The QTPP can include features such as intended clinical use, route of administration, dosage form and strength, delivery system, container/closure system, attributes affecting PK/PD, and other key product quality criteria (such as sterility, purity, and stability) [2].

2.2.2 Critical Quality Attributes (CQAs)

The core objective of the QbD approach is to determine the Critical Quality Attributes (CQAs) of the product under development. CQAs can be defined as physical, chemical, biological, or microbiological properties or characteristics that should be within an appropriate limit, range, or distribution to ensure the desired product quality [1]. A summary of the numerous analytical techniques used to test and quantify CQAs is provided in Section 2.3.

In order to identify the CQAs, risk assessment methodologies are used to sort the quality attributes identified in the QTPP along a criticality continuum based on their impact on safety and efficacy. Examples of such methodologies include Risk Ranking, Preliminary Hazards Analysis (PHA), and Failure Mode and Effects Analysis (FMEA) [3–5]. These methodologies generally involve assigning a numerical evaluation to risk as a function of severity (or impact), likelihood (or uncertainty), and/or detectability.

2.2.3 Critical Process Parameters (CPPs) and Design Space

Any process parameter whose variability affects a CQA is deemed a Critical Process Parameter (CPP). CPPs should thus be monitored or controlled to ensure the process produces the desired quality [1]. Process characterization is the in-depth investigation to determine how the CPPs affect the CQAs and culminates in the Design Space. Given the complexity of the manufacturing process for a biologic therapeutic, the Design Space is typically a multi-dimensional combination and interaction of CPPs and other input variables (i.e., material attributes) that assure that CQAs will remain within specifications. Process characterization makes use of statistical techniques such as Design of Experiments (DOE), Multivariate Data Analysis (MVDA), and Statistical Process Control (SPC) [6]. Qualified small-scale models can be extremely helpful to establish the Design Space, as they enable a more thorough testing [7].

Process Analytical Technology (PAT) can play a crucial role during process characterization, especially if technologies enabling real-time continuous data collection are developed and/or implemented, as they can reveal the complex relationships between CQAs and CPPs, contributing to a more comprehensive process understanding and ultimately a more effective control.

2.2.4 Control Strategy

The scientific understanding of the sources of variability of CQAs generated by the process characterization lays the foundation to determine a planned set of controls to ensure process performance (in terms of quality). This set of controls is referred to as the Control Strategy in QbD terminology. The Control Strategy can include parameters and attributes related to the drug substance or drug product, along with the entire manufacturing process, equipment, and facility operating settings, in-process controls, finished product specifications, and the associated methods and frequency of monitoring and control [1].

The US-FDA defines Process Validation as the collection and evaluation of data, from the process design stage through commercial production, which establishes scientific evidence that a process is capable of consistently delivering quality products [8]. In its recommendations for Process Validation, the FDA establishes three stages within a lifecycle approach to accomplish this objective: Process Design (definition of the process through development and scale-up), Process Qualification (evaluation of the process design to determine if the process is capable of reproducible commercial manufacturing), and Continued Process Verification (to demonstrate that the process remains in a state of control) [8]. The Control Strategy is a key outcome of the process design stage and thus supports the efforts of process validation/qualification, which will be ultimately required for market approval.

2.3 OVERVIEW OF THE ANALYTICAL TECHNOLOGIES IN THE BIOLOGICS CQAs

The analytical technologies described in Section 2.3 translate QbD concepts into measurable evidence by linking CQAs to actionable data across unit operations. The following subsections summarize the principles and practical PAT deployment of key analytical families, indicating where methods are applied (upstream, downstream, or both) and how they support real-time monitoring or fit within at-/off-line control strategies. The subclasses of techniques and tools are grouped into eight classes of analytical approaches, with multiple subclasses within each class.

2.3.1 Chromatography-Based Methods

Principle: Chromatography separates components of a complex biological mixture based on differences in their interaction with a stationary phase and a mobile phase.

The interacting factors could be charge-based, size-based, hydrophobicity-based, or affinity-based, to name a few key factors that depend on the target biomolecule. In size exclusion chromatography (SEC), biomolecules are separated by apparent hydrodynamic size as they diffuse through pores in the stationary phase: larger species elute earlier, smaller ones later. In ion-exchange chromatography (IEC or IEX), proteins and other biomolecules are resolved according to net surface charge, which determines how strongly they bind to charged ligands on the resin and at what salt or pH they elute. Hydrophobic interaction chromatography (HIC) exploits differences in surface hydrophobicity, where more hydrophobic molecules interact more strongly with mildly hydrophobic ligands at high salt. Affinity chromatography (AC) uses highly specific interactions—such as antibody–antigen or protein A–Fc binding—to capture a target while impurities flow through. By coupling these separation mechanisms to spectroscopic or mass spectrometric detectors, chromatography enables precise quantification and characterization of product variants and impurities that define CQAs. Table 2.1 presents a list of commonly used chromatography techniques for CQA assessment.

TABLE 2.1
Different Chromatography Approaches and Their Relevance to CQAs in Biologics

Sub-Technique	LOD/Sensitivity	Upstream/ Downstream	PAT Mode	Application
SEC–UV	1–10 µg/mL	Downstream	At-line/off-line	Aggregation, size variants
SEC–MALS	10–50-µg load	Downstream	At-line/off-line	Mass, aggregates
HIC–UV	Low-µg/mL; hydrophobic variants ≥1%	Downstream	At-line/off-line	ADC drug-to-antibody ratio, hydrophobically modified species
Affinity (Protein A etc.)–UV	Low-µg/mL; dynamic range 0.05–10 g/L	Both (harvest + DS)	At-line/off-line	Titer measurement (mAbs), capture pool purity, leached ligand monitoring
2D-LC/multi-D LC–MS	ng–low-µg on-column; ppm–ppb impurities	Downstream	At-line/off-line (rarely on-line)	Deep heterogeneity mapping (sequence variants, PTMs, HCPs)
UHPLC for glycans (HILIC-FLD/MS)	~pmol–nmol sugar per injection	Downstream	At-line/off-line	N-glycan profiles, sialylation/galactosylation (mAbs, Fc-fusions)
Capillary electrophoresis (CE-SDS, cIEF, ZipChip CE–MS)	~0.1–1 µg/mL protein; charge isoforms ≥0.1–1%	Both	At-line/off-line	Purity, size, charge isoforms, quick titer, and quality snapshots from bioreactors
AEX/SEC for AAV (UV/FL/MS)	~10^{11}–10^{12} vg/mL; empty/full discrimination by area ratios	Both	At-line/off-line	Empty/full capsid ratio, vector purity, co-eluting impurities ~0.1–1 µg/mL protein; charge isoforms ≥0.1–1%

Examples of PAT applications in biologics using chromatography-based methods: Chromatography remains the workhorse for CQA measurement in biologics, with SEC, IEX, HIC, and affinity modalities frequently coupled to optical, fluorescence, or mass-spectrometric detectors for detailed characterization of product and process-related variants [9]. In particular, SEC coupled with multi-angle light scattering (SEC-MALS) established the principle of absolute molar mass and aggregate quantification without the need for calibration standards. It has been extensively applied to monoclonal antibodies (mAbs) for rapid, at-line aggregation quality check during development and comparability studies [10]. Under typical conditions, SEC-MALS can quantify aggregate populations on the order of ~0.1–0.5% of total protein, depending on injected mass and optical configuration [10].

Recent analytical advances have pushed chromatographic separations toward online multi-dimensional analysis. For example, four-dimensional workflows such as HIC×SEC-ion mobility×mass spectrometry (MS) or SEC×SEC-IM×MS that deconvolute size and conformational heterogeneity of IgGs and antibody–drug conjugates into minor variants while maintaining native-like conditions. These methods are currently configured as online or at-line hyphenated systems, rather than fully inline PAT, because of the constraints imposed by solvent compositions, flow rates, and MS interfaces. Charge-variant analysis by IEX chromatography—especially cation-exchange chromatography (CEX)—coupled online to native high-resolution MS has clarified the early "first-use" principles for charge-variant identification of intact mAbs. More recently, direct hyphenation of CEX to collision-induced unfolding (CIU) in ion-mobility MS has been used to interrogate subtle conformational differences among charge isoforms that cannot be resolved by UV detection alone, providing a structurally sensitive, at-line PAT tool for charge and higher-order structure monitoring [11].

In upstream bioprocessing, rapid at-line capillary electrophoresis (CE) or UHPLC methods, including microfluidic CE–MS platforms such as ZipChip, have been deployed to monitor titer and glycosylation with cycle times on the order of minutes, thereby enabling more intensive process characterization and feed-forward control strategies [9]. In downstream chromatography skids, at-line UHPLC for charge variants and N-glycans, sometimes integrated into two-dimensional LC–MS workflows, is increasingly used to infer pool composition and trigger column changeovers or adjust pooling windows, effectively acting as an at-line PAT layer superimposed on process-scale UV monitoring [12, 13].

For vaccines and viral vectors, affinity capture—such as protein A or tailored affinity ligands—combined with fast multi-dimensional LC–MS is used to characterize impurities (e.g., host cell proteins (HCPs)) and product variants in cell-culture harvests or intermediate pools [12]. SEC and anion-exchange (AEX) chromatography for adeno-associated virus (AAV) are commonly coupled to UV, fluorescence, or MS detectors at-line to estimate empty and full capsid ratios, a critical CQA in gene-therapy products [14]. In terms of detection limits, SEC-UV typically detects proteins in the low-µg/mL range, while SEC-MALS generally requires tens of micrograms of injected protein for robust molar-mass fitting [10]. Native IEX–MS methods can operate with sub-microgram injections, and state-of-the-art Orbitrap or time-of-flight instruments routinely achieve ng-level on-column sensitivity for variant detection, provided that desalting and chromatographic conditions are optimized [15]. Overall, chromatography-based PAT for biologics remains predominantly at-line or on-line (using closed sampling loops to external analyzers) in development and pilot-scale settings, with fully in-line integration largely confined to optical concentration monitors or spectroscopic surrogates during column elution.

2.3.2 Gel-Based Methods

Principle: Gel-based methods separate biomolecules by size, charge, or a combination of both within a hydrated polymer matrix. In SDS-PAGE, proteins are denatured by sodium dodecyl sulphate (SDS), which imparts a net negative charge and linearizes the amino acid chain. When an electric field is applied, separation occurs primarily by molecular size as smaller proteins migrate more rapidly through the gel. Contrary to SDS-PAGE, Native PAGE preserves the protein's native conformation and charge, so migration reflects a combination of size, shape, protein conformation, and structural complexity, and net charge, allowing oligomeric states or complexes to be visualized. In two-dimensional electrophoresis, proteins are first separated by isoelectric point (IEF) and then by size, increasing resolving power. After separation, proteins are visualized through stains or detected by antibodies (Western blot), enabling identification or semi-quantitative assessment of specific species. These methods provide a "fingerprint" of a sample's molecular composition and are especially useful for confirming identity, detecting degradation fragments, and revealing gross changes in heterogeneity, including in upstream processes for locating the product in a cell, as a soluble fraction, membrane fraction, or inclusion bodies.

Examples of use of gel-based methods in PAT: SDS-PAGE and native PAGE, along with two-dimensional PAGE and Western blotting, remain foundational techniques for visualizing size and charge heterogeneity in biologics and are still widely used for process characterization, impurity profiling, and identity/purity confirmation, albeit exclusively as off-line PAT tools. Modern fluorescent stains such as SYPRO Ruby or related SYPRO formulations, together with optimized silver staining protocols, have pushed the sensitivity of gel-based methods into nanogram or even sub-nanogram range per band, enabling semi-quantitative assessment of trace impurities in enriched fractions [16, 17]. For HCPs, silver stain and SYPRO Ruby can visualize as low as nanogram amounts per band, but they lack the breadth of coverage and absolute sensitivity now attainable by immunoassays or LC–MS–based HCP workflows [13]. Semi-quantitative densitometry of total protein or specific bands across lots is still useful for impurity tracking in small R&D setups, while Western blots extend specificity to particular proteins—such as proteases in enzyme products or residual structural proteins in viral vectors—through antibody-based detection that can reach sub-nanogram sensitivity with optimized chemiluminescent substrates [16]. Table 2.2 provides a list of common applications of gel-based techniques applied to CQAs.

In subunit vaccine programs, native PAGE and Blue-Native PAGE provide insight into oligomerization and assembly states—such as trimeric vs. monomeric antigen forms—where changes in band patterns can be correlated with immunogenicity or stability. For AAV and other viral vectors, native gels can distinguish fully assembled

TABLE 2.2
Different Gel-and Staining-Based Approaches and Their Relevance to CQAs in Biologics Manufacturing

Sub-Technique	LOD	Upstream/ Downstream	PAT Mode	Application
SDS-PAGE	50–100 ng	Both	Off-line	Purity, fragments
Silver/SYPRO	0.1–5 ng	Both	Off-line	Trace impurities
Western blot	Sub-ng	Both	Off-line	Specific identity
2D-PAGE (IEF + SDS)	~1–10 ng/spot	Mainly downstream	Off-line	High-resolution impurity and variant mapping in development

capsids from subassemblies or degraded forms, thereby supporting PD and comparability. However, all of these gel-based approaches are fundamentally batchwise and off-line, which restricts their use to orthogonal confirmation and comparability rather than real-time or closed-loop control. Reported limits of detection for modern fluorescent stains are typically 0.25–1 ng/band for SYPRO Ruby, 0.1–0.25 ng/band for optimized silver staining, and ~4–8 ng/band for SYPRO variants, depending on imaging conditions [16, 17]. Despite these sensitivities, assay-to-assay variability, differences in stain binding across proteins, and effects of sample preparation constrain their use as primary release assays or quantitative PAT metrics.

Consequently, gel-based methods continue to anchor orthogonal characterization for clone selection, scale-up, and hold-condition development, acting as rapid off-line "quality checks" that complement faster at-line separations and label-free sensing technologies (described below). Recent reviews of analytical platforms for mAbs and related biologics consistently position gels and Western blots as confirmatory or diagnostic tools alongside chromatographic, spectrometric, and immunochemical methods, rather than as direct inputs to automated control strategies [9, 12] (Figure 2.2).

2.3.3 Nucleic Acids–Based Methods

Principle: Nucleic acid–based methods exploit the specificity of base pairing and the exponential amplification capabilities of polymerases to detect and quantify DNA or RNA. In qPCR, a defined DNA sequence is amplified through cyclic temperature changes, and the accumulation of product is monitored in real time using fluorescent dyes or probes whose signal increases with the amount of amplified DNA. The cycle at which fluorescence exceeds a threshold is inversely related to the initial template concentration, enabling sensitive quantitation over several orders of magnitude. Digital PCR partitions a sample into thousands of microreactors, such that each partition contains either zero or one (or a few) template molecules; after amplification, counting positive vs. negative partitions yields an absolute copy number based on Poisson statistics. For RNA, reverse transcription creates complementary DNA (cDNA) that is then amplified in the same way. These principles enable sensitive

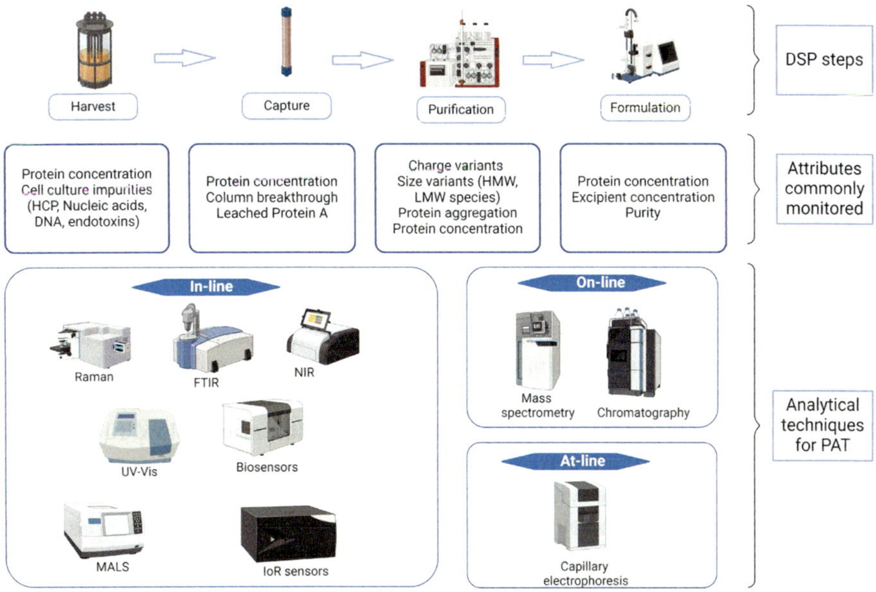

FIGURE 2.2 Principles and use of different PAT tools for QbD approach. (Figures reproduced from [18].)

measurement of residual host-cell DNA, viral genomes, plasmid templates, and therapeutic mRNA, making nucleic acid methods central to controlling genetic impurities and potency-linked nucleic acid attributes. A list of commonly used nucleic acid-based methods applied to CQA monitoring is presented in Table 2.3.

Examples of use of nucleic acid-based methods in PAT: Nucleic acid–based assays, particularly quantitative PCR (qPCR) and digital PCR (ddPCR), underpin

TABLE 2.3
Different Nucleic Acid Based Approaches and Their Relevance to CQAs in Biologics Manufacturing

Sub-Technique	LOD	Upstream/ Downstream	PAT Mode	Application
qPCR	1–10 pg DNA	Both	At-line	Residual DNA
ddPCR	Single-copy	Both	At-line	Viral genome titers
RT-qPCR	Single-digit copies	Upstream	At-line	mRNA integrity
RT-ddPCR	Single-copy sensitivity for transcripts	Upstream (IVT) and potency	At-line/ off-line	mRNA copy number, transgene expression in transduction assays, potency markers
UV 260 nm (A_{260})	~0.1–0.5 µg/mL nucleic acid	Both	At-line/ Off-line	Bulk DNA/RNA concentration (crude estimation, often combined with qPCR/ddPCR)

on hydrodynamic size and molar-mass distributions, enabling detection of nano- to micron-scale aggregates and shifts in particle size distributions that may be tied to stability or potency [24]. Typical detection thresholds are modality-dependent: DLS in backscatter configuration can characterize hydrodynamic radii from roughly 1–10 nm up to ~1 μm at low-μg/mL protein concentrations, whereas MALS generally requires tens of micrograms injected to provide precise molar-mass estimates [10].

2.3.5 Protein Interaction Assays

Principle: Protein interaction assays (see Table 2.5) measure the binding between a molecule of interest (e.g., a therapeutic antibody or antigen) and its specific partner (receptor, ligand, or antibody) to infer functional activity. In label-free techniques such as surface plasmon resonance (SPR) and bio-layer interferometry (BLI), one binding partner is immobilized on a sensor surface, and changes in the local refractive index or interference pattern are monitored as the other partner associates and dissociates from the surface. These optical changes are proportional to the mass bound at the surface and are recorded as sensorgrams over time. By fitting these curves, the association and dissociation rate constants and the equilibrium dissociation constant can be extracted, yielding a kinetic and affinity profile. In classical immunoassays like ELISA, binding events are translated into a colorimetric, fluorescent, or chemiluminescent signal via enzyme-linked secondary antibodies. In all cases, the fundamental principle is that specific molecular recognition events are converted into measurable physical signals whose magnitude and kinetics report on the strength, specificity, and abundance of the interaction.

Examples of use of protein interaction-based methods in PAT: Protein interaction assays, especially label-free biosensors such as SPR and BLI, have become indispensable for linking CQAs to functional binding properties. These technologies were first widely adopted in bioprocess development to obtain real-time kinetic parameters for antibody–antigen and receptor–ligand interactions, and they now support potency and identity testing for mAbs, subunit vaccines, and complex biologics. In a PAT context, SPR and BLI are currently configured primarily as at-line or near-line systems due to the need for sample clarification, control of nonspecific binding, and mitigation of sensor fouling. Nevertheless, their kinetic parameters—association

TABLE 2.5
Different Protein Interaction Assays and Their Relevance to CQAs in Biologics Manufacturing

Sub-Technique	LOD	Use	Mode	Application
SPR	ng/mL	DS	At-line	Binding kinetics
BLI	ng/mL	DS	At-line	Potency, Fc binding
ELISA (sandwich/competitive)	pg–ng/mL	Both	At-line	Antigen potency
Lab-on-chip, ELISA	Similar to ELISA, shorter analysis	Both	At-line	Rapid potency and impurity checks with small sample volumes

rate (k_a), dissociation rate (k_d), and equilibrium dissociation constant (K_D)—map directly onto mechanistically relevant CQAs such as antigen binding, receptor occupancy, or Fc effector engagement.

During formulation and fill-finish of vaccines and protein therapeutics, BLI or SPR can be used to confirm that antigen–adjuvant adsorption or excipient changes have not compromised epitope availability or binding activity. For Fc-mediated biologics, BLI assays measuring binding to Fcγ receptors or neonatal Fc receptor (FcRn) can serve as surrogates for effector function and in vivo exposure, while enzyme biologics may be monitored via binding to cofactors or inhibitors as a functional proxy. For mRNA-based therapeutics and vaccines, BLI has recently been adapted to detect double-stranded RNA impurities—an emerging CQA related to innate immune activation—by capturing dsRNA on specialized surfaces and measuring binding in real time [22]. Typical limits of detection for SPR/BLI in well-optimized assays fall in the low-ng/mL to sub-ng/mL range for high-affinity interactions, enabling sensitive detection of changes in active-binding fraction.

ELISA and other immunoassays remain complementary, high-specificity interaction tools, generally implemented as off-line or at-line assays with turnaround times of tens of minutes to a few hours. Modern microfluidic immunoassay systems and high-throughput BLI platforms allow multi-analyte potency and impurity profiling within development or QC labs, and FDA potency-assay guidance increasingly emphasizes the importance of mechanism of action–aligned binding or cell-based assays (CBAs) for novel mAbs and therapeutic proteins [3, 25]. Overall, protein interaction assays are not yet in-line PAT tools, but their proximity to the process and direct alignment with clinical mechanisms make them critical anchors for potency, stability, and comparability assessments.

2.3.6 Light-Based Methods

Principle: Light-based methods (common examples listed in Table 2.6) probe biomolecules and particles by measuring their absorption, emission, scattering, or modification of light. Vibrational spectroscopies such as Raman and IR detect characteristic molecular vibrations: in IR, molecules absorb light at frequencies corresponding to vibrational transitions; in Raman, inelastic scattering shifts the frequency of incident photons according to vibrational energy levels. These spectra provide "chemical fingerprints" that encode information about molecular composition and structure. In

TABLE 2.6
Different Scattering Methods and Their Relevance to CQAs in Biologics Manufacturing

Sub-Technique	LOD	Use	Mode	Application
Raman	mM–sub-mM	US	In-line	Nutrients, control
FTIR/NIR	mg/mL	Both	In-line	Concentration
DLS	Low μg/mL	DS	At-line	Nanoparticle size

NIR and FTIR, broad overtone and combination bands can be correlated with concentration and composition via multivariate calibration. Scattering techniques like DLS analyze fluctuations in scattered light intensity caused by Brownian motion, allowing hydrodynamic size distributions to be inferred from diffusion coefficients. Focused Beam Reflectance Measurement (FBRM) measures chord lengths of particles crossing a scanned laser spot, generating real-time size distributions in turbid slurries, while particle imaging (particle vision monitoring (PVM), micro-flow imaging (MFI)) directly captures images of particles and uses image analysis to determine size and morphology. Across these methods, light–matter interactions are translated into spectral or spatial signals that reflect composition, size, structure, or aggregation state.

Examples of light-based methods in PAT: Light-based methods—including vibrational spectroscopy, UV–Vis/NIR/IR spectroscopy, scattering, and imaging—provide rapid, minimally invasive PAT readouts across upstream and downstream unit operations. Raman spectroscopy has emerged as a leading in-line probe for upstream control, capable of simultaneously quantifying glucose, glutamine, lactate, ammonia, and total cell density via multivariate models. Case studies in CHO fed-batch systems have demonstrated successful closed-loop control of glucose and lactate using in situ Raman analysis, leading to improved cell density, productivity, and even modulation of mAb glycosylation profiles when combined with appropriate feeding strategies [26]. These methods are genuinely in line: spectra are acquired through optical probes inserted into the bioreactor, and chemometric models—sometimes maintained using real-time or just-in-time learning frameworks—provide continuous predictions of key metabolites and CQAs.

Near-infrared (NIR) and mid-IR/FTIR spectroscopy are widely used for raw-material identification and have been applied to in-line monitoring of concentration, osmolality, and formulation composition for proteins and excipients, including as surrogates for HPLC during chromatography elution; here, overlapping spectra necessitate robust multivariate calibration [27]. In the context of particle attributes for vaccines, adjuvants, and LNPs, scattering-based methods such as DLS, laser diffraction (LD), and FBRM, together with imaging systems like PVM, have been deployed to monitor hydrodynamic size, chord-length distributions, and morphology. For aluminum-adjuvanted vaccines, combined off-line LD and in-line FBRM have been used to track precipitated $AlPO_4$ particle size during formation and antigen adsorption, revealing that dilution and degree of protein adsorption significantly influence particle size distributions—and, potentially, immunogenicity [24, 27].

MFI and related microflow digital imaging (MDI) approaches extend light-based analytics to the sub-visible particle range (~2–100 μm), enabling the detection and morphological classification of proteinaceous aggregates, silicone oil droplets, and other contaminants in both clear and opaque formulations, including aluminum-adjuvanted vaccines [28]. These at-line methods have been widely adopted as quality guardrails during formulation, filtration, and stress-testing studies for parenteral biologics. Detection limits for light-based methods are strongly modality-specific: Raman typically quantifies metabolites in the sub-mM to low-mM range, with prediction errors governed by calibration data; FTIR/NIR concentration monitors detect proteins in the low-mg/mL range in flow; DLS resolves nanometer-scale shifts at low-µg/mL protein levels; FBRM measures chord-length distributions from roughly 1 to 1000 µm in-line; and MFI images particles as small as ~2 µm, with overall sensitivity defined by sample volume and optical configuration [26, 28].

2.3.7 Enzymatic Assays

Principle: Enzymatic assays (see Table 2.7 for some examples) quantify biochemical activity by coupling substrate conversion to a measurable signal. An enzyme catalyzes the transformation of a specific substrate into a product under defined conditions, typically following Michaelis–Menten kinetics. In many assays, the substrate or product is designed to be chromogenic, fluorogenic, or luminescent, so that the rate or extent of signal change is directly proportional to enzymatic activity. In some cases, the enzyme acts in a cascade, where its product becomes the substrate for another reaction that amplifies the signal. By comparing the rate of signal development to that of a standard, the activity of a therapeutic enzyme, viral neuraminidase (NA), or reporter system can be quantified. The principle thus links molecular function—catalytic turnover—to a robust optical readout, providing a functional measure of potency or inhibition that complements purely structural analytics.

Examples of use of enzymatic assays in PAT: Enzymatic assays are archetypal potency assays for enzyme biologics and also play important roles in vaccines and mAbs, where catalytic activity relates directly or indirectly to the clinical mechanism. Their core principle—measuring substrate turnover under defined conditions—translates into off-line or at-line formats that are fast, sensitive, and often highly stability-indicating. For coagulation factor products such as Factor VIII, validated chromogenic assays quantify activity with high precision and are widely regarded as gold-standard potency methods, supporting both labeling and stability testing [29]. These assays typically achieve limits of detection in the low-picomolar range for FVIII activity and have been carefully standardized across reagent kits and instruments to minimize inter-laboratory variability [30].

In influenza vaccines, NA enzymatic activity serves as an important functional marker and potential correlate of breadth of protection. Chemiluminescent assays such as NA-STAR and fluorescent MUNANA-based assays measure NA activity and inhibition with high sensitivity and have been extensively used for monitoring virus susceptibility to NA inhibitors, screening candidate antivirals, and characterizing vaccine-induced antibodies [31]. Optimized NA-STAR assays demonstrate

TABLE 2.7
Different Enzymatic Approaches and Their Relevance to CQAs in Biologics Manufacturing

Sub-Technique	LOD	Use	Mode	Application
Chromogenic FVIII	pM–nM	DS	Off-line	Potency
NA activity	ng/mL	DS	Off-line	Vaccine potency
ADCC reporter	ng/mL	DS	At-line	Fc potency

limits of quantitation in the low-ng/mL range for native NA, with dynamic ranges suitable for both research and quality-control settings. For therapeutic mAbs whose mechanisms involve Fc-mediated effector functions, cell-coupled enzymatic reporters—such as ADCC reporter gene assays—quantify FcγR engagement and downstream signaling, and have been validated as potency assays for lot release, stability testing, and biosimilarity assessments [25].

These enzymatic and reporter-based assays are typically implemented off-line or at-line, with assay times ranging from under an hour (e.g., some chromogenic or luminescent assays) to several hours for complex cell-based reporter systems. True in-line enzyme activity sensors are rare in GMP biologics manufacturing because of matrix complexity, specificity requirements, and the need for controlled assay conditions. Nonetheless, enzymatic potency assays are irreplaceable for function-linked CQAs and can serve as critical feedback signals in process characterization, hold-time studies, excipient screening, and formulation optimization, even if they are not used directly in closed-loop control [3, 29, 32].

2.3.8 Cell-Based Methods

Principle: Cell-based methods assess the biological activity of a product within living cells, integrating receptor binding, signal transduction, internalization, trafficking, and downstream functional responses (see common examples listed on Table 2.8). Typically, target cells are exposed to the biologic under controlled conditions, and a specific cellular response is measured—such as reporter gene activation, biomarker secretion, cell killing, or rescue of a disease phenotype. In reporter assays, ligand–receptor engagement triggers a signaling cascade that leads to expression of a luciferase or fluorescent protein, converting cellular signaling into a quantifiable optical signal. In transduction assays for viral vectors, successful delivery and expression of a transgene are measured by reporter expression or functional correction. The central principle is that the complex cellular machinery acts as the sensor: the magnitude and shape of the response curve reflect the biologic's effective potency and mechanism in a system that more closely mimics in vivo conditions than purely biochemical assays.

Examples of the use of cell-based assays in PAT: CBAs are widely recognized as the gold-standard potency assays for many biologics because they integrate binding, signaling, internalization, and downstream biological responses into a

TABLE 2.8
Different Cell-Based Approaches and Their Relevance to CQAs in Biologics Manufacturing

Sub-Technique	LOD	Use	Mode	Application
ADCC CBA	10–100 ng/mL	DS	Off-line	Functional potency
Reporter assays	Low-ng/mL	DS	At-line	Mechanism potency
AAV transduction	10^3–10^4 vg/mL	DS	Off-line	Vector potency

single functional readout. Their early widespread use in bioprocess analytics centered on antiviral and antitumor mAbs—through ADCC and antibody-dependent cellular phagocytosis (ADCP) reporter gene assays—and on vaccine components for which cell-based readouts of antigen function (e.g., NA or HA activity) were relevant [25, 33].

For viral vectors, cell-based transduction potency assays are now central to CQA frameworks. Quantitative AAV potency assays that measure both transgene expression and biological activity in transduced cells have been developed and validated, and are increasingly used to support comparability, stability, and release, often with readouts based on reporter expression, restored enzyme activity, or specific functional endpoints [34]. Recent work has focused on engineering cell lines with enhanced susceptibility to AAV serotypes and on tailoring promoters to improve in vitro transduction efficiency and the dynamic range of potency assays [35]. In enzyme replacement therapies and gene-therapy approaches for lysosomal storage disorders, CBAs that measure cellular rescue of disease phenotypes—such as clearance of accumulated substrates or normalization of lysosomal function—are critical, even when analysis times are longer, because they directly reflect the therapeutic mechanism [36].

Although CBAs exhibit higher variability and longer cycle times than physicochemical PAT tools, they uniquely anchor "what matters clinically" and thus are essential orthogonal potency assays in QbD strategies for mAbs, enzymes, viral vectors, and emerging gene and cell therapies. Automation, miniaturization, and microfluidic multiplexing are gradually moving CBAs closer to at-line implementation, but they will likely remain off-line for the foreseeable future in routine GMP manufacturing [37].

2.3.9 Other Methods

A suite of complementary tools fills gaps left by classic separations, spectroscopy, and CBAs, particularly in biomass monitoring and complex particle characterization. A few examples are listed in Table 2.9. Dielectric spectroscopy (bio-capacitance) measures the frequency-dependent permittivity of cell suspensions. It has become a standard in-line estimator of viable cell density and biomass in mammalian and microbial bioreactors. Its industrial "first uses" involved installing capacitance probes directly in stainless-steel and single-use bioreactors to track viable cell

TABLE 2.9
Different Hybrid Methods and Approaches, Their Relevance to CQAs in Biologics Manufacturing

Sub-Technique	LOD	Use	Mode	Application
Dielectric capacitance	Viable volume ~1–2%	US	In-line	VCD
wNMR	Formulation-level	DS	At-line	Adjuvant state
FBRM/PVM	Size 1–1000 μm	DS	In-line	Particles

volume in real time; over the last three decades, the technology has matured into multi-frequency dielectric spectroscopy capable of distinguishing viability changes and physiological states [38–40]. The underlying principle—that viable cells with intact membranes behave as microscopic capacitors in an alternating electric field—enables continuous, label-free monitoring of viable biomass, and capacitance signals are now routinely integrated into feed, perfusion, aeration, and harvest control algorithms [39].

In particle analytics for vaccines and LNPs, in-line FBRM and in situ imaging tools (PVM) provide critical information on the evolution of size and morphology during precipitation, adsorption, and emulsification steps. Studies of aluminum-adjuvanted vaccines have used FBRM to monitor particle size distribution in situ during $AlPO_4$ precipitation and subsequent antigen adsorption. At the same time, PVM has given visual confirmation of morphological changes or aggregation events that could compromise potency [27]. Weak-field nuclear magnetic resonance (wNMR) has recently emerged as a non-invasive, at-line method for characterizing aluminum-adjuvanted formulations; it has been shown to outperform LD in differentiating high- and low-dose vaccine formulations and to provide complementary information on sedimentation kinetics and the compactness of settled vaccines.

Imaging-based particle counting using MFI or related flow microscopy systems provides sub-visible particle counts and morphological classification for protein biologics and vaccines, including formulations containing opaque adjuvants [28]. Particle size ranges from ~1 to 300 μm, with lower detection limits governed by optical configuration and sampling volume. These methods are predominantly at-line, as they require sample withdrawal into imaging systems, but they are now standard for parenteral quality and for investigating process-induced particle formation during filtration, pumping, or agitation. Collectively, dielectric spectroscopy (in-line), FBRM/PVM (in-line), wNMR (at-line), and MFI/MDI (at-line) sit naturally alongside spectroscopic and chromatographic PAT tools within multivariate monitoring and control strategies, expanding process visibility into viable biomass, particle attributes, and formulation behavior in ways that more traditional methods cannot.

2.4 CONCLUSION

Techniques that are both in-line and online, as well as offline, support the objectives of QbD for biologics manufacturing processes. There is a general trend to increasingly implement in-line and on-line techniques into continuous monitoring for quality attributes, as there is a strong emphasis on data during production processes in pharma and biologics manufacturing. In this regard, advanced label-free biosensing approaches are developing rapidly to meet the technology push in the biomanufacturing community. This chapter has focused on all the attributes of monitoring, rather than on a specific online, offline, or in-line approach, to provide a comprehensive outlook on this topic, while situating the monitoring of CQAs in the overall perspective of QbD, and the role that PAT can have in supporting the QbD endeavor.

REFERENCES

1. ICH Harmonized Tripartite Guideline, "Pharmaceutical development," *Q8 (2R). Revised August,* 2009.
2. L. X. Yu *et al.*, "Understanding pharmaceutical quality by design," *The AAPS Journal*, vol. 16, no. 4, pp. 771–783, 2014.
3. ICH Guideline, "Quality Risk Management," Q9 (R1) *Adopted January*, 2023.
4. C. B. W. Group, "A-Mab: a case study in bioprocess development," *Emerville: CASSS*, 2009.
5. M. F. Witcher, "Failure mode and effect analysis (FMEA) as a quality by design (QbD) tool for managing biopharmaceutical product development and manufacturing risks," *Bioprocessing Journal*, vol. 13, no. 3, 2014.
6. P. Pokhriyal, P. Gupta, S. Khambhampaty, R. Ullanat, and M. Pathak, "Recent trends in data analytics for upstream process workflows," 2022.
7. C. Finkler and L. Krummen, "Introduction to the application of QbD principles for the development of monoclonal antibodies," *Biologicals*, vol. 44, no. 5, pp. 282–290, 2016.
8. (2011). *FDA-2008-D-0559, Process Validation: General Principles and Practices* [Online] Available: https://www.fda.gov/regulatory-information/search-fda-guidance-documents/process-validation-general-principles-and-practices
9. H. A. Alhazmi and M. Albratty, "Analytical techniques for the characterization and quantification of monoclonal antibodies," *Pharmaceuticals*, vol. 16, no. 2, p. 291, 2023.
10. G. Brusotti, E. Calleri, R. Colombo, G. Massolini, F. Rinaldi, and C. Temporini, "Advances on size exclusion chromatography and applications on the analysis of protein biopharmaceuticals and protein aggregates: a mini review," *Chromatographia*, vol. 81, no. 1, pp. 3–23, 2018.
11. G. van Schaick, E. Domínguez-Vega, J. Castel, M. Wuhrer, O. Hernandez-Alba, and S. Cianférani, "Online collision-induced unfolding of therapeutic monoclonal antibody glyco-variants through direct hyphenation of cation exchange chromatography with native ion mobility–mass spectrometry," *Analytical Chemistry*, vol. 95, no. 8, pp. 3932–3939, 2023.
12. J. Guo, R. Kufer, D. Li, S. Wohlrab, M. Greenwood-Goodwin, and F. Yang, "Technical advancement and practical considerations of LC-MS/MS-based methods for host cell protein identification and quantitation to support process development," *mAbs*, 2023, vol. 15, no. 1, pp. 2213365.
13. S. Hessmann, C. Chery, A.-S. Sikora, A. Gervais, and C. Carapito, "Host cell protein quantification workflow using optimized standards combined with data-independent acquisition mass spectrometry," *Journal of Pharmaceutical Analysis*, vol. 13, no. 5, pp. 494–502, 2023.
14. D. Dobnik *et al.*, "Accurate quantification and characterization of adeno-associated viral vectors," *Frontiers in Microbiology*, vol. 10, p. 1570, 2019.
15. R. Sadighi *et al.*, "Online multimethod platform for comprehensive characterization of monoclonal antibodies in cell culture fluid from a single sample injection-Intact protein workflow," *Analytica Chimica Acta*, vol. 1287, p. 342074, 2024.
16. L. A. Beer and D. W. Speicher, "Protein detection in gels using fixation," *Current Protocols in Protein Science*, vol. 91, no. 1, pp. 10–5. 1-10.5. 20, 2018.
17. T. F. Scientific, "SYPRO Ruby Protein Gel Stain," ed, 2007.
18. P. Sathiyapriyan *et al.*, "Current PAT landscape in the downstream processing of biopharmaceuticals," *Analytical Science Advances*, vol. 6, no. 1, p. e70013, 2025.
19. H. Yang, "Establishing acceptable limits of residual DNA," *PDA Journal of Pharmaceutical Science and Technology*, vol. 67, no. 2, pp. 155–163, 2013.

20. J. Lebron *et al.*, "Adaptation of the WHO guideline for residual DNA in parenteral vaccines produced on continuous cell lines to a limit for oral vaccines," *Developments in Biologicals*, vol. 123, pp. 35–44; discussion 55, 2006.
21. Y. Wang, N. Menon, S. Shen, M. Feschenko, and S. Bergelson, "A qPCR method for AAV genome titer with ddPCR-level of accuracy and precision," *Molecular Therapy Methods & Clinical Development*, vol. 19, pp. 341–346, 2020.
22. D. S. Silas *et al.*, "Development of biolayer interferometry (BLI)-based double-stranded RNA detection method with application in mRNA-based therapeutics and vaccines," *Pharmaceutics*, vol. 16, no. 9, p. 1227, 2024.
23. B. Rajendar, M. J. Reddy, A. Madhavi, G. Shravankumar, G. S. Rao, and R. V. Matur, "Multi angle light scattering as a process analytical technology tool for real-time monitoring of molar mass of protein-polysaccharide conjugate fractions," *Journal of Chromatography Open*, vol. 2, p. 100045, 2022.
24. E. Shardlow, M. Mold, and C. Exley, "From stock bottle to vaccine: elucidating the particle size distributions of aluminum adjuvants using dynamic light scattering," *Frontiers in Chemistry*, vol. 4, p. 48, 2017.
25. B. S. Parekh *et al.*, "Development and validation of an antibody-dependent cell-mediated cytotoxicity-reporter gene assay," *mAbs*, 2012, vol. 4, no. 3, pp. 310–318.
26. S. Craven, J. Whelan, and B. Glennon, "Glucose concentration control of a fed-batch mammalian cell bioprocess using a nonlinear model predictive controller," *Journal of Process Control*, vol. 24, no. 4, pp. 344–357, 2014.
27. C. Mei *et al.*, "Aluminum phosphate vaccine adjuvant: analysis of composition and size using off-line and in-line tools," *Computational and Structural Biotechnology Journal*, vol. 17, pp. 1184–1194, 2019.
28. D. K. Sharma, D. King, P. Oma, and C. Merchant, "Micro-flow imaging: flow microscopy applied to sub-visible particulate analysis in protein formulations," *The AAPS Journal*, vol. 12, no. 3, pp. 455–464, 2010.
29. S. Butenas, B. Parhami-Seren, M. Gissel, E. Gomperts, D. Fass, and K. Mann, "Potency and mass of factor VIII in FVIII products," *Haemophilia*, vol. 15, no. 1, pp. 63–72, 2009.
30. R. A. Marlar, K. Strandberg, M. Shima, and D. M. Adcock, "Clinical utility and impact of the use of the chromogenic vs one-stage factor activity assays in haemophilia A and B," *European Journal of Haematology*, vol. 104, no. 1, pp. 3–14, 2020.
31. R. C. Buxton, B. Edwards, R. R. Juo, J. C. Voyta, M. Tisdale, and R. C. Bethell, "Development of a sensitive chemiluminescent neuraminidase assay for the determination of influenza virus susceptibility to zanamivir," *Analytical Biochemistry*, vol. 280, no. 2, pp. 291–300, 2000.
32. G. S. Padickakunnel and N. V. Gupta, "Modern FDA guidance and comparative overview of FDA and EMA on process validation," *International Journal of Pharmacy and Pharmaceutical Sciences,* vol. *6, no. 4*, pp. 14–17, 2014.
33. C. Liu *et al.*, "Development of a novel reporter gene assay to evaluate antibody-dependent cellular phagocytosis for anti-CD20 therapeutic antibodies," *International Immunopharmacology*, vol. 100, p. 108112, 2021.
34. S. J. Aronson *et al.*, "A quantitative in vitro potency assay for adeno-associated virus vectors encoding for the UGT1A1 transgene," *Molecular Therapy Methods & Clinical Development*, vol. 18, pp. 250–258, 2020.
35. J. Zengel, E. S. Esterman, A. Ponnuswami, N. R. Wall, and J. E. Carette, "Development of cell lines with increased susceptibility to diverse adeno-associated viral vectors to enable in vitro potency assays," *Molecular Therapy Methods & Clinical Development*, vol. 33, no. 1, 2025.
36. J. H. LeBowitz, J. H. Grubb, J. A. Maga, D. H. Schmiel, C. Vogler, and W. S. Sly, "Glycosylation-independent targeting enhances enzyme delivery to lysosomes and

decreases storage in mucopolysaccharidosis type VII mice," *Proceedings of the National Academy of Sciences*, vol. 101, no. 9, pp. 3083–3088, 2004.

37. P. Salmikangas, B. Carlsson, C. Klumb, T. Reimer, and S. Thirstrup, "Potency testing of cell and gene therapy products," *Frontiers in Medicine*, vol. 10, p. 1190016, 2023.
38. W.-J. Xu, Y. Lin, C.-L. Mi, J.-Y. Pang, and T.-Y. Wang, "Progress in fed-batch culture for recombinant protein production in CHO cells," *Applied Microbiology and Biotechnology*, vol. 107, no. 4, pp. 1063–1075, 2023.
39. S. Metze, S. Ruhl, G. Greller, C. Grimm, and J. Scholz, "Monitoring online biomass with a capacitance sensor during scale-up of industrially relevant CHO cell culture fed-batch processes in single-use bioreactors," *Bioprocess and Biosystems Engineering*, vol. 43, no. 2, pp. 193–205, 2020.
40. C. Justice *et al.*, "Process control in cell culture technology using dielectric spectroscopy," *Biotechnology Advances*, vol. 29, no. 4, pp. 391–401, 2011.

3 PAT Tools for Process Monitoring of Biologics Manufacturing

Ayyappasamy Sudalaiyadum Perumal and Malvika Nagrath

3.1 WORKFLOW OF THE PROCESS ANALYTICAL TECHNOLOGY FRAMEWORK FOR BIOLOGICS

To integrate Process Analytical Technology (PAT) into the development and production of biologics, critical process parameters (CPPs) and critical quality attributes (CQAs) must first be identified to establish the desired product profile [1, 2], subsequently characterized through process monitoring to gain a comprehensive understanding of how process inputs influence product quality [2]. Such understanding supports early decision-making in the biomanufacturing process design and layout [1, 2].

The characterization of CPPs and CQAs through process monitoring requires a detailed understanding of the biologic's profile and typically progresses through the following steps (Figure 3.1a) [1, 2]:

1. ***Recognition of need***: Define the information required from process monitoring for the specific biologic profile. What parameter needs characterization? Is it a contaminant, an intermediate product, the final product, or a by-product?
2. ***Selection of monitoring approach***: Determine the most appropriate strategy to obtain the required information. For instance, will monitoring the metabolic state, cellular state, media state, or fluorophore-associated compounds capture the necessary data?
3. ***Definition of performance criteria***: Establish the performance requirements of the selected approach. Should the method prioritize robustness, repeatability, sensitivity, accuracy, or a combination of these attributes?
4. ***Analytical tool selection***: Choose the analytical tool (e.g., capacitance, fluorometry, and spectroscopy) that best meets the defined approach and performance criteria.
5. ***Calibration and validation***: Calibrate the chosen analytical tool to ensure accurate, reliable, and consistent process monitoring.

3.2 APPROACHES TO PROCESS MONITORING

Process monitoring in biomanufacturing can be conducted by analyzing the **cell culture environment**, the **cell culture process**, and/or the **cell state** for upstream processing and analyzing **particle size, purification**, and so on for downstream

 DOI: 10.1201/9781032662961-3

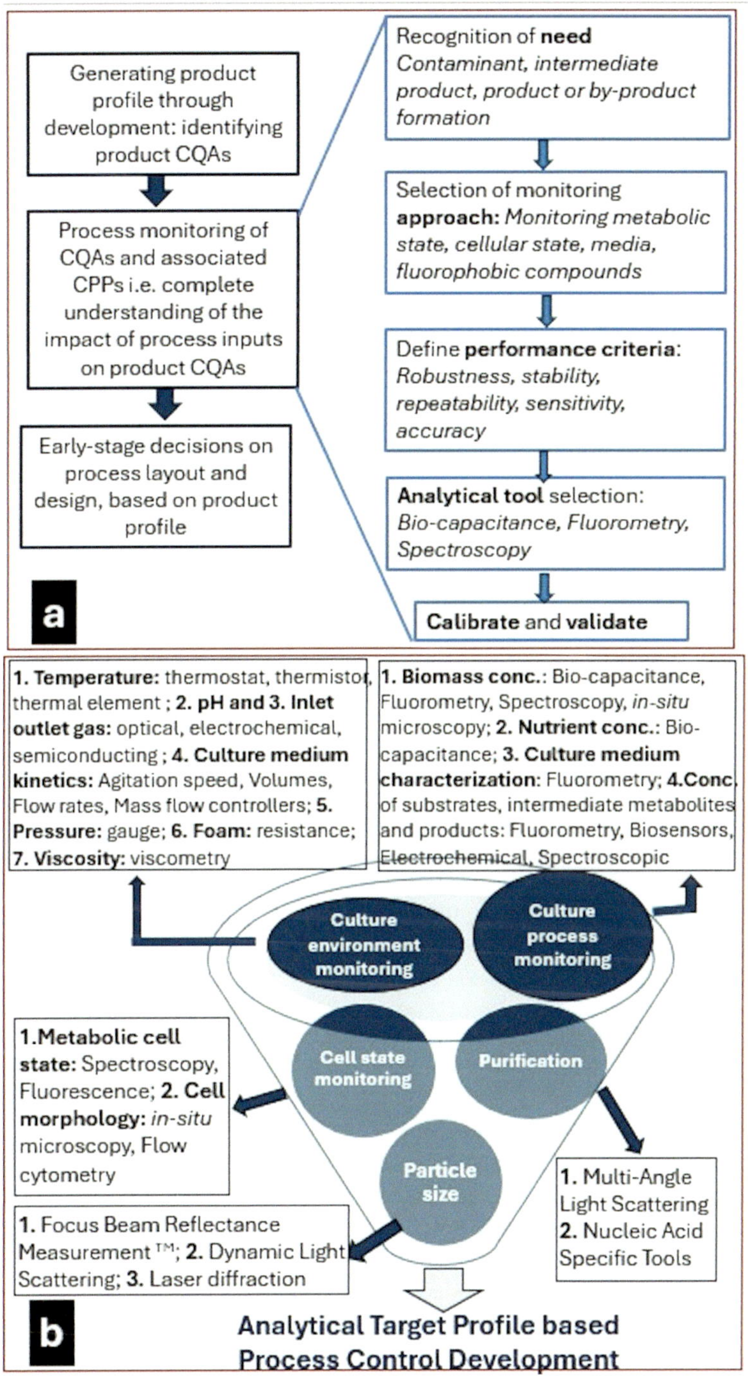

FIGURE 3.1 (a) Process monitoring workflow of the PAT framework for biologics; (b) approaches for process monitoring.

processing. Optimization of these parameters is essential for maintaining consistent product quality and achieving higher yields. High and consistent productivity is attainable only when cells in a *stable physiological state* are cultured under *optimized environmental conditions* using *efficient culture processes*, supported by robust process control strategies (Figure 3.1b). The **cell state** is typically assessed through monitoring of cell morphology and metabolic activity. **Cell culture environment monitoring** includes parameters such as temperature, pH, dissolved oxygen (DO) and carbon dioxide, medium kinetics, pressure, foam formation, and viscosity. The importance of these variables in cell culture processes is well established, and a wide range of electrochemical and optical sensors is available for their measurement. To optimize the **cell culture process**, monitoring of biomass, nutrients, substrates, intermediate metabolites, and final product concentrations is required. Analytical tools such as bio-capacitance, fluorometry, spectroscopy, in situ microscopy, and biosensors are commonly employed for this purpose.

By no means does the above description encompass an exhaustive list of approaches involved in the whole biomanufacturing process. There are other procedural steps in the *downstream processing*, such as recovery, clarification, purification, concentration, and polishing, that require analyzing particle size, so on and so forth. A few examples are captured in Figure 3.1b.

A variety of analytical tools and sensors are currently used in biomanufacturing to generate **real-time data** for process monitoring [3, 4] at low frequency [3, 4]. In such cases, **soft sensors** are employed. Soft sensors rely on mathematical models, informed by one or more sensor signals, to estimate otherwise difficult-to-measure variables [3, 4]. For example, an in situ near-infrared (NIR) sensor used to monitor biomass concentration can be applied to estimate the specific growth rate from the NIR signal [4].

Recent research has focused on the development of **multi-sensor chips** for monitoring key parameters in biomanufacturing [5]. In parallel, **integrated sensor platforms** are being designed to capture real-time data and transmit signals remotely using wireless communication [6].

3.3 TOOLS FOR PAT

3.3.1 Physico-Chemical Tools

Physico-chemical tools form the core of PAT in both traditional pharmaceutical manufacturing [7, 8] and modern Bioprocessing [9], like in the production of biologics [10] like enzymes, proteins, antibodies, acids and alcohols, cell culture–derived products [11], and gene therapy products (DNA and mRNA biologics) [12]. These tools measure fundamental physical and chemical process parameters that can be directly correlated with the product quality, process performance, and compliance with regulatory expectations under *Quality by Design (QbD) [13]* principles (Chapter 2). In PAT, physico-chemical tools often serve as the first layer of process monitoring, providing continuous, real-time data on parameters, such as temperature, pH, conductivity, osmolarity, turbidity, viscosity, particle size, and surface charge, particularly in the upstream, like during the bioreactor stage. They are typically robust, simple to integrate, and compatible with a range of process formats, from at-line testing to fully in-line monitoring.

Physico-chemical measurements are often considered "basic" process parameters, but within a PAT framework, they gain additional value when multivariate analysis links them to CQAs and CPPs. When combined with chemometric models, these seemingly routine measurements can serve as predictive indicators of product performance, stability, and safety [14]. These measurements, although often low in analytical complexity, are highly valuable because they can be acquired continuously [14] and interpreted in real-time [14], even remotely [15], thanks to the modern signal monitoring capacity. Furthermore, many of these sensing approaches mentioned above find applications in other stages of biomanufacturing, like during downstream processing [16, 17], fill, and finish steps [18]. For instance, conductivity measurements, pH, and ionic concentration measurements are key during chromatography approaches [19, 20]. Dedicated sensors in the chromatography approaches constantly measure the column and instrument parameters to achieve a reproducible purification step. The multiple sensing approaches for different PAT aspects are organized below as principles, applications, advantages, and limitations.

3.3.1.1 Core Physico-Chemical Measurement Parameters in PAT

Physico-chemical PAT tools operate at the juncture of well-established principles of process engineering and analytical chemistry. There are wide tools and technologies that aid in highly scalable measurements from micro to mini research and development (R&D)-specific bioreactors to large-scale high-production-capacity reactors. The following are the key parameters that are well integrated in the bioreactor stage.

3.3.1.1.1 Temperature Sensors

Temperature measurement, for instance, is achieved through thermocouples or resistance temperature detectors (RTDs). Thermocouples [21] and RTDs provide kinetic control information, important for reactions and cell culture growth. These sensors provide accurate readings of kinetic control variables, which are essential for ensuring the correct folding of proteins during expression or the stability of biomolecules during storage. A deviation of even a few degrees can result in altered glycosylation patterns in monoclonal antibodies or loss of conformational epitopes in protein-based vaccines.

3.3.1.1.2 pH Electrodes

These probes measure hydrogen ion activity, crucial in fermentation, protein stability, and crystallization control. pH measurement, typically achieved through glass electrode probes, plays a similarly critical role. The hydrogen ion activity directly affects enzyme activity, protein solubility, and nucleic acid stability, influencing product potency and shelf-life. In the production of recombinant proteins, pH shifts during fermentation can trigger stress responses in host cells, reducing yields or increasing the formation of unwanted variants [22].

3.3.1.1.3 Viscosity Sensors

They monitor rheological changes in slurries, gels, or high-density cell cultures, including in recent times during plasmid purification relevant to gene therapy applications (mRNA technology [23]. Viscosity sensors monitor rheological changes in slurries, gels, or high-density cell cultures, providing early warning of process deviations such as over-concentration during ultrafiltration/diafiltration (UF/DF) or improper mixing in formulation tanks. Though these measurements are often considered low in analytical complexity, their ability to be acquired continuously and interpreted in real time makes them invaluable for PAT-enabled control strategies [8, 24].

3.3.1.1.4 Other Physico-Chemical Measurements

- **Conductivity meters** measure ionic strength, useful for monitoring buffer composition, desalting steps, or chromatography elution profiles.
- **Osmometers** measure osmolarity, influencing cell viability in mammalian cell cultures.
- **Turbidity meters** (e.g., optical density (OD) probes) estimate biomass or particulate load during fermentation or filtration [25].
- **DO probes** [19, 26]

3.3.1.2 Common Physico-Chemical PAT Technology in Bioprocessing

3.3.1.2.1 Focused Beam Reflectance Measurement

Focused Beam Reflectance Measurement (FBRM) uses a rotating laser beam that scans across suspended particles in the process stream. The instrument records backscattered light signals to determine chord length distribution, which serves as a surrogate for particle size [27].

In **vaccine manufacturing**, FBRM has been used to monitor antigen–adjuvant adsorption and detect aggregation events that may reduce potency. In vaccine manufacturing, PAT tools such as ReactIR, ParticleTrack with FBRM, and EasyViewer were used to monitor real-time adsorption of tetanus toxoid (TT) to aluminum phosphate ($AlPO_4$) and aluminum oxyhydroxide (AlOOH) adjuvants. In situ probes tracked conformational changes, particle size distributions (PSDs), and aggregation kinetics, enabling visualization of multi-step adsorption mechanisms. This upstream application allowed optimization of antigen–adjuvant binding and early detection of aggregation events that could compromise vaccine potency [28]. Kumar et al. applied analytical PAT-like assays to evaluate the compatibility of recombinant rotavirus vaccine (t-NRRV) antigens with aluminum-adjuvanted pediatric pentavalent formulations. Competitive immunoassays monitored antigen–adjuvant binding stability, while adsorption/desorption studies identified destabilizing factors such as specific antigens, preservatives, and mixed adjuvant systems. Upstream formulation adjustments, including switching to an AH-only adjuvant, mitigated antigen desorption and preserved stability, providing insights for multidose combination vaccine development [29].

In **crystallization processes**, it tracks particle size growth and can trigger process adjustments to prevent oversized particles [30, 31]. A novel PAT-integrated framework combines FBRM with artificial neural networks (ANNs) to convert chord length distributions into crystal size distributions in real time. Using ANN-driven modeling, nucleation and growth kinetic parameters were estimated directly during crystallization, enabling dynamic control of supersaturation and PSD. This enhances predictive monitoring, reduces dependence on off-line particle analysis, and supports upstream crystallization optimization in batch and semi-batch processes [32]. FBRM, attenuated total reflectance–Fourier-transform infrared (ATR–FTIR), and UV/Vis spectroscopy are reviewed as PAT tools for monitoring protein crystallization in downstream purification. Real-time FBRM tracks nucleation events, crystal growth, and size distribution, while inversion models convert chord length data into accurate PSDs for low-aspect-ratio protein crystals like lysozyme, recombinant insulin, and vitamin B12. Integration with machine learning-based image analysis further improves monitoring of morphology, polymorphism, and aggregation, enabling process control and better product quality [30]. In a continuous reactive crystallization system for active pharmaceutical ingredient (API) production, PAT tools, including inline ReactIR and FBRM, were embedded in a plug-flow reactor-continuous stirred tank reactor (PFR–CSTR) cascade. ReactIR measured reactant concentrations, while FBRM tracked chord length distributions to control nucleation and optimize growth. The system achieved stable particle sizes (~150 μm) and improved crystallization yield by ~2.7%, demonstrating PAT's effectiveness in **continuous upstream manufacturing** and supporting integrated continuous manufacturing (ICM) strategies [31]. This review highlights PAT-driven strategies for controlling PSD (PSD) in pharmaceutical crystallization. Inline tools such as FBRM and ReactIR enable real-time monitoring of nucleation and growth kinetics, supporting optimization of cooling profiles, seeding protocols, and agitation rates. Case studies show that adaptive control using FBRM can maintain supersaturation within the metastable zone, minimizing fines and ensuring crystals meet critical size specifications. These approaches are widely applied in **upstream crystallization** stages to ensure downstream processability. Real-time, non-destructive works in opaque systems [30] provide chord length rather than true particle diameter, which requires calibration for specific particle morphologies [33].

3.3.1.2.2 Laser Diffraction (LD)

Laser diffraction (LD) is another key technology in particle sizing techniques, operating by measuring the angular scattering pattern of laser light as particles pass through a laser beam in a suspension like cell culture solutions, biomass, solute crystallization, and formulation set-ups (precipitates) [34, 35]. Kowal et al. [34] reported the application of multiple PAT probes, including FBRM, infrared spectroscopy, turbidity, and conductivity sensors, to monitor aluminum phosphate ($AlPO_4$) adjuvant manufacturing in real time. The study demonstrated how these tools enabled continuous tracking of particle formation, size distribution, and reaction kinetics during upstream processing. Integration of multi-probe PAT facilitated optimization of mixing conditions, pH adjustment, and reactant flow, ensuring consistent product quality and improved process control. Using Mie or Fraunhofer theory, the scattering pattern is converted into a PSD.

Off-line LD has been used for **vaccine adjuvant characterization** at raw material, intermediate, and final stages. Off-line LD was employed to characterize aluminum phosphate ($AlPO_4$) vaccine adjuvant at raw-material, intermediate, and final stages. The study compared PSDs measured off-line (LD) with in-line FBRM and found consistent results across scales and stages—highlighting stable PSDs throughout manufacturing. This off-line LD provided essential batch consistency data and supported quality control in **upstream adjuvant production**, complementing in-line PAT tools for comprehensive process oversight [36]. Established technique with well-developed data interpretation models [37]. Usually off-line; sample preparation may introduce variability.

3.3.1.2.3 Turbidity and Optical Density Probes

Turbidity and OD probes function by quantifying the attenuation of light caused by suspended particles in the medium [19, 38, 39]. These measurements are crucial for continuous biomass monitoring in fermenters, where they support dynamic feeding strategies in fed-batch or continuous processes. For example, in *Escherichia coli* fermentation for recombinant protein production, OD measurements guide glucose feeding to avoid overflow metabolism and acetate accumulation. In downstream processing, turbidity probes can detect breakthrough during depth filtration, signaling the need for filter changeover before contamination of the filtrate occurs. Measure light attenuation caused by suspended particles [40].

Continuous biomass monitoring in fermenters to optimize feed strategies [41]. **Zanzotto et al. [41]** developed a microscale membrane-aerated microbioreactor integrating OD, DO, and pH sensors for continuous monitoring of *E. coli* cultures. Their system tracked biomass and physiological state in real time, replicating growth kinetics seen at bench scale. This facilitated high-throughput screening and precise characterization of upstream fermentation dynamics. Comprehensive PAT infrastructures for recombinant *E. coli* cultivation, incorporating in-line, on-line, and at-line measurements (e.g., substrate, product, cell density) linked through Supervisory Control Data Acquisition (SCADA) and software tools like MATLAB for real-time control and optimization of feeding strategies in high-density upstream fermentations have been developed.

Detection of filter breakthrough during depth filtration. Researchers applied statistical enhancement of laser turbidity readings to greatly increase sensitivity for early detection of particulate breakthrough in filtration systems. By analyzing fluctuations in laser nephelometric turbidity with simplified real-time algorithms, they demonstrated improved identification of filter compromise even when particle concentrations were low—offering a proactive PAT approach to maintain downstream product clarity (**downstream filtration monitoring**). **Thakur et al. [17]** engineered a continuous depth-filtration skid equipped with turbidity and pressure sensors to automate dead-end filtration processes. This PAT deployment enabled real-time detection of filtration breakthrough and system fouling, facilitating timely interventions and ensuring continuous process robustness in **downstream clarification** operations. It is inexpensive, simple, and easily integrated into control systems; also, it is non-specific—cannot differentiate between cell types or debris without additional analytics [38].

3.3.2 Biomolecular Tools for PAT

Biomolecular PAT tools are analytical platforms specifically designed to monitor **biologically derived molecules**—including proteins, nucleic acids, polysaccharides, lipids, and their complexes—throughout the biomanufacturing lifecycle. While physico-chemical tools assess general process conditions, biomolecular tools provide **molecule-specific information**, enabling the detection of changes in **structure, conformation, function, and interactions** that directly influence the CQAs of biopharmaceuticals or biologics [10]. These tools are essential in biologics and vaccine manufacturing because they allow for the direct monitoring of CQAs such as antigen structure, functional activity, and aggregation state, all of which directly impact product efficacy and safety.

Traditional biomolecular assays like enzyme-linked immunosorbent assays (ELISA), Western blotting, or high-performance liquid chromatography (HPLC) have long been used for such purposes [42]. These techniques are often off-line, labor-intensive, and not easily adaptable to real-time monitoring. The evolution of PAT has pushed the development of rapid, non-destructive, and ideally in-line tools that can perform molecularly specific measurements without compromising process sterility or speed. The operating principles of biomolecular PAT tools are rooted in specificity-driven detection.

In the context of biologics and vaccines, biomolecular tools are crucial for the following:

1. Monitoring antigen structure and stability (relevant for monoclonal antibody, mAb, and other polyclonal antibody production processes). Measuring binding interactions between biomolecules (e.g., antigen–adjuvant, receptor–ligand) for determining the efficacy of the therapeutic targets [43].
2. Assessing protein folding, aggregation, and degradation relevant for enzymes, hormones, and other therapeutic proteins production process [44].
3. Tracking nucleic acid purity and integrity in mRNA, DNA, or viral vector products [45].
4. By detecting process-related impurities such as host cell proteins (HCPs) or residual DNA, for example, ELISA-type assays for HCP and double-stranded RNA in mRNA biologics [46], Nilsson et al. (1994) implemented a **flow-through ELISA** system directly connected to fermentation and purification modules, enabling **real-time quantification of product concentration** during processing. This online configuration streamlined sample handling—by coupling bioreactor outflow with immediate immunoassay detection—thereby enabling **continuous upstream monitoring** that linked fermentation performance with purification readiness and ensured tighter process control. Additionally, in recent times, for the detection of the double-stranded RNA in mRNA biologics, and during the quality control of the in vitro mRNA synthesis, ELISA-based assays were used. These are mostly off-line assays, but rapid, and can be tested in the form of lateral flow assays to improve speed and point of biologics manufacturing applications [47].

3.3.2.1 Principles of Biomolecular PAT Measurement

Biomolecular PAT tools operate using **specificity-driven detection mechanisms**, often based on the following:

1. Molecular recognition (e.g., antibody–antigen binding).
2. Spectroscopic signatures of biomolecules (e.g., intrinsic protein fluorescence).

3. Light scattering to infer macromolecular size and aggregation.
4. Label-free biosensing using refractive index or mass change at a sensor surface.

3.3.2.2 Common Biomolecular Tools for PAT Technology in Bioprocessing

3.3.2.2.1 Dynamic Light Scattering (DLS)

Dynamic Light Scattering (DLS) is widely used to measure PSDs of submicron particles by detecting fluctuations in scattered light caused by Brownian motion [48]. In Bioprocessing, DLS is commonly applied to monitor antigen particle size before and after adjuvant adsorption [49]. For instance, in aluminum-adjuvanted vaccines, the size distribution of antigen–adjuvant complexes can be correlated with immunogenicity [50]. DLS can also assess protein stability during accelerated stability studies; for example, detecting early aggregation in SARS-CoV-2 spike protein formulations stored at 4°C. Online DLS prototypes are emerging that allow real-time particle sizing during formulation, enabling immediate intervention if aggregates form [51]. It also found key use in mRNA technology, for determining the lipid nanoparticle (LNP) particle sizes and polydispersity index [52, 53]. DLS measures fluctuations in scattered light intensity caused by Brownian motion of particles in solution, relating this motion to hydrodynamic radius using the Stokes–Einstein equation [54]. Sensitive to submicron particles, rapid measurement is applicable in opaque media with a backscatter configuration. It requires dilution for high-concentration samples, and polydisperse systems may complicate data interpretation. Temperature has been also identified as a critical parameter to control variability [55].

Reiterating similar applications, from above, DLS is used for multiple biomanufacturing PAT applications.

1. Monitoring antigen size distribution before and after adjuvant adsorption [43].
2. Assessing protein stability during stress testing (e.g., SARS-CoV-2 spike protein stability at 4°C).
3. Online DLS prototypes for real-time particle sizing of protein antigens and oil-in-water adjuvants. LNP formulation studies use DLS approaches, a key QbD and quality control tool for the size distribution of the mRNA LNPs. Additionally, DLS is a key tool in viral vector-based vaccine manufacturing related to QbD. DLS is used for viral vector size measurements [56].

3.3.2.2.2 Multi-Angle Light Scattering (MALS)

Multi-Angle Light Scattering (MALS) measures the intensity of light scattered at multiple angles, allowing for the determination of absolute molecular weight and size distribution without the need for calibration standards [57]. When coupled with size-exclusion chromatography (SEC-MALS), it provides a powerful tool for monitoring protein molar mass, detecting aggregation, and verifying product purity during downstream processing. For example, in monoclonal antibody manufacturing, in-line MALS can identify the onset of dimer or higher-order aggregate formation during polishing steps, allowing for process adjustments in real time [58]. Measures scattered light intensity at multiple angles to determine **absolute molecular weight** and **size distribution** of particles or macromolecules. No need for calibration standards as it gives the direct molecular weight. **Limitations:** Sensitive to dust and bubbles, and requires optical clarity.

3.3.2.2.3 Bio-Layer Interferometry (BLI) and Surface Plasmon Resonance (SPR)

Bio-Layer Interferometry (BLI) and Surface Plasmon Resonance (SPR) are label-free biosensing techniques capable of quantifying molecular interactions in real time. BLI measures changes in interference patterns as molecules bind to immobilized ligands on a fiber-optic sensor, whereas SPR detects changes in refractive index near a gold-coated surface. Both are used extensively in vaccine and therapeutic protein development to characterize antigen–antibody or receptor–ligand interactions. For example, during influenza vaccine formulation, SPR can verify that hemagglutinin retains its receptor-binding activity after adjuvant adsorption. Although typically applied at-line due to sensor fouling risks in crude fluids, these tools are invaluable for process development and CQA verification. BLI measures interference pattern changes as molecules bind to immobilized ligands on a fiber-optic biosensor. **SPR** detects refractive index changes near a metal film as molecules associate/dissociate from the surface. It also provides real-time monitoring of **antigen–antibody** or **ligand–receptor** interactions, process development screening for **adjuvant–antigen binding kinetics**, label-free and quantitative binding kinetics, which are typically at-line sensor fouling in crude process fluids.

3.3.2.2.4 Immunochemical Assays Adapted for PAT

Immunochemical assays, such as ELISA, remain among the most specific methods for detecting proteins, HCPs, or process-related impurities. Recent developments have produced portable ELISA-on-chip devices capable of at-line HCP quantification in minutes rather than hours. Such rapid testing can be used during purification to ensure that residual HCP levels meet specifications before advancing to the next step, reducing the risk of batch reprocessing or rejection [59]. It uses antigen–antibody recognition for specific detection, often coupled with colorimetric, chemiluminescent, or fluorescence readouts, rapid HCP quantification using portable ELISA-on-chip devices [60], and at-line antigen potency testing during vaccine fill–finish [60]. High specificity and sensitivity, still largely off-line, require reagents and calibration standards.

3.3.2.2.5 Nucleic Acid–Specific Tools

Nucleic acid–specific assays are increasingly important with the rise of gene therapies and mRNA-based vaccines [11]. Fluorescence-based quantification methods, such as PicoGreen assays, allow for sensitive detection of double-stranded DNA contaminants in viral vector preparations. UV absorbance at 260 nm is a traditional method for quantifying nucleic acids, and when integrated into flow-through cuvettes, it can be adapted for at-line monitoring of mRNA concentration during LNP manufacturing. More advanced techniques, such as capillary electrophoresis with fluorescence detection, can assess mRNA integrity to ensure translational competence.

3.3.3 Dielectric Properties Measurement Tools for PAT

Dielectric property measurement tools are a specialized category of PAT instruments designed to monitor the electrical properties of biological materials and process fluids [61, 62]. Their most common application in Bioprocessing is real-time measurement of viable cell density (VCD) in cell culture and fermentation systems [63]. This is particularly valuable in mammalian cell culture for therapeutic protein production, insect cell culture for viral vector manufacturing, and yeast or microbial fermentation for recombinant protein expression [63, 64].

The fundamental principle behind dielectric spectroscopy is that viable cells with intact plasma membranes behave differently from non-viable cells when exposed to an alternating electric field. The cell membrane acts as a capacitor, storing and releasing charge in a frequency-dependent manner. By measuring the complex permittivity of a sample over a range of frequencies, it is possible to distinguish viable from non-viable cells and estimate parameters such as capacitance (related to viable biomass) and conductivity (related to ionic strength of the medium) [65]. In mammalian cell bioreactors producing monoclonal antibodies, dielectric probes can be used to monitor growth phase transitions and trigger nutrient feeds at the optimal time to maintain productivity while preventing metabolite accumulation. In microbial fermentation, dielectric measurements can detect the onset of the stationary phase, informing harvest timing for maximum yield [66].

Dielectric spectroscopy can be implemented via in-line probes inserted directly into the bioreactor or through at-line flow-through sensors. Multi-frequency systems provide richer data by detecting shifts in characteristic frequency that can indicate changes in cell size, membrane composition, or stress responses. These tools integrate readily into feedback control loops, enabling fully automated feeding and harvest strategies.

Dielectric property measurement tools are a specialized class of PAT instruments designed to probe the electrical characteristics of biological materials and process fluids. In Bioprocessing, these measurements are most commonly used to monitor cell culture viability and biomass concentration in real time, particularly in mammalian, insect, yeast, and microbial fermentation systems.

Dielectric tools are highly relevant to PAT because of the following:

1. They enable continuous, real-time measurements without sampling.
2. They provide direct correlation to CQAs such as productivity (often linked to VCD) and product quality.
3. They integrate readily into feedback control loops for nutrient feed, aeration, and harvest timing.

3.3.3.1 Principles of Dielectric Measurement in PAT

Dielectric properties describe how a material interacts with an electric field and are typically represented by: permittivity (ε')—the material's ability to store electric energy, and dielectric loss (ε'')—the material's tendency to dissipate energy as heat. When an alternating current (AC) electric field is applied across a suspension of cells, viable cells with intact membranes act as tiny capacitors, causing a frequency-dependent polarization effect. Non-viable cells or debris lack membrane integrity and do not exhibit the same capacitive behavior. By scanning over a range of frequencies (typically 0.1–20 MHz), the dielectric spectrum can be analyzed to determine: capacitance (proportional to the volume fraction of viable cells), conductivity (related to the ionic strength of the medium), and characteristic frequency (linked to cell size and membrane properties).

3.3.3.2 Common Dielectric PAT Tools for PAT Technology in Bioprocessing

In-Line Dielectric Probes: They are installed directly in the bioreactor vessel through standard ports. Measure capacitance and conductivity in real time. Output can be scaled to give viable cell volume (VCV) or VCD in units such as cells/mL or g dry weight/L—for example, Aber Instruments' Futura probes, Hamilton Incyte sensors.

At-Line Flow-Through Sensors: A sample is drawn continuously or intermittently through a measurement cell outside the reactor. It reduces fouling risk but requires pumping systems and sterilization protocols.

Multi-Frequency Dielectric Spectroscopy: It extends beyond single-frequency capacitance to acquire full dielectric spectra and enables discrimination of cell subpopulations (e.g., viable vs. stressed) based on membrane integrity and composition.

Dielectric Tools Are Primarily Used to Monitor: VCD, real-time tracking of culture growth phase; cell size changes, characteristic frequency shifts may indicate swelling (e.g., nutrient depletion) or shrinkage (e.g., apoptosis); and physiological transitions, drop in capacitance may signal onset of stationary or death phase, supporting timely harvest.

3.3.4 Spectroscopy-Based Tools

Among the various available techniques, spectroscopy, particularly vibrational spectroscopy, is highly suited for in-line monitoring [67]. Spectroscopy analyzes the interaction of electromagnetic radiation with matter, producing spectra that select molecular composition and structure. Vibrational spectroscopy is especially valuable due to its rapid, non-destructive nature and minimal sample preparation.

3.3.4.1 Principle: Vibrational Spectroscopy

When electromagnetic radiation interacts with molecules, it can induce vibrational transitions such as bending, stretching, and rocking. These transitions occur because molecular bonds can absorb or transmit specific wavelengths of radiation corresponding to their vibrational energy levels [68–70].

Electromagnetic radiation can be described either as waves or particles (photons), each photon possessing a discrete amount of energy [69]. Atoms within a molecule are bound by quantized vibrational energy states. In equilibrium, a bond exists in its lowest vibrational energy level, denoted by the vibrational quantum number, $v = 0$. Absorption of electromagnetic radiation energy can excite the bond to a higher vibrational level, such as $v = 1$, termed as the fundamental transition. This *fundamental transition* is the primary basis of MIR spectroscopy [69]. Additional energy absorptions can excite the bond to $v = 2,3,4$ and higher levels; these are known as overtone transitions, which form the basis of NIR spectroscopy [69]. The vibrational energy levels ($v = 0,1,2,3,\ldots$) are equally spaced [69].

Molecules containing multiple atoms can exhibit simultaneous active vibrations, producing combination bands. Like overtones, these have weaker absorbance than fundamental transitions [68, 69]. For example, the asymmetric O–H stretch in water at ~3750 cm^{-1} and the O–H bending mode at ~1630 cm^{-1} combine to produce a band at ~5380 cm^{-1} (3750 + 1630).

Three main types of vibrational spectroscopy are typically used in PAT: NIR, MIR, and Raman spectroscopy. While MIR and NIR spectroscopies are based on the absorption of electromagnetic radiation, Raman spectroscopy operates on the principle of light scattering. The selection of a specific technique depends on the biological characteristics, the production environment, and the analytical goal.

3.3.4.2 Common Spectroscopy-Based Tools

3.3.4.2.1 Mid-Infrared (MIR) Spectroscopy

MIR spectroscopy involves measuring the absorption or transmission of MIR radiation between 2.5 and 50 μm (4000 and 400 cm^{-1}) by a sample [68, 69]. When MIR light interacts with a molecule, specific wavelengths are absorbed, causing covalent bonds within the molecule to stretch, bend, or twist. A spectrometer captures this data, generating an MIR spectrum—a plot of absorbance or transmittance vs. wavenumber (the inverse of wavelength) [69]. Various statistical tools such as second derivative analysis can be applied to the acquired spectra to gain specific information needed for quantification analysis [69].

MIR spectroscopy is particularly sensitive to functional groups that are distinct structural motifs within molecules that consistently absorb MIR radiation at characteristic frequencies [69]. For instance, the carbonyl group (C = O) in a ketone absorbs MIR radiation around a specific wavenumber, regardless of the molecule's overall structure. This property enables MIR spectroscopy to correlate spectral features (peaks) with molecular structure [68, 69]. These peaks are like the fingerprints of specific bonds or molecules; the reference tables for the peaks are extensively available in the literature.

Organic molecules absorb light in the MIR range at specific wavelengths, producing their fingerprints on the MIR spectrum. Taking advantage of this feature, MIR spectroscopy can provide valuable insights into molecular-level changes such as protein denaturation, aggregation, metabolite identification, protein concentration, and post-translational modifications like phosphorylation and glycosylation during biological manufacturing [71]. These analyses help optimize process parameters and ensure biologic stability and functionality [71].

3.3.4.2.1.1 Challenges and Integration in Biomanufacturing Despite its potential, the application of MIR spectroscopy in real-time biologic manufacturing remains limited. This is largely due to challenges such as water interference in spectra and the complexity of integrating MIR equipment into manufacturing lines. However, technological advancements—such as fiber optic probes, ATR sensors, and automatic background subtraction—are mitigating these issues, enabling more robust in-line spectral acquisition and analysis [67].

Biologics present specific manufacturing challenges [72]:

1. Heat sensitivity.
2. Susceptibility to contamination.
3. Multi-step, intricate processing.

3.3.4.2.1.2 Advantages These challenges contribute to risks like protein aggregation and misfolding [71]. MIR spectroscopy, particularly when implemented in-line, enables real-time monitoring of protein structure and stability, helping prevent these issues [71]. Absorbance spectra are preferred for quantitative analysis, as per Beer–Lambert's Law, which states that absorbance is directly proportional to concentration [69].

3.3.4.2.1.3 Protein Structure and MIR Spectroscopy Proteins, the key components of most biologics, have a hierarchical structure [73]:

1. ***Primary structure***: A linear sequence of amino acids joined by peptide bonds (Figure 3.2a). Amino acids each contain a central carbon atom, a carboxyl group (–COOH), an amino group (–NH_2), and a unique R-group that defines its chemical behavior (Figure 3.2b). Peptide bonds, formed between the amino group of one amino acid and the carboxyl group of another, are

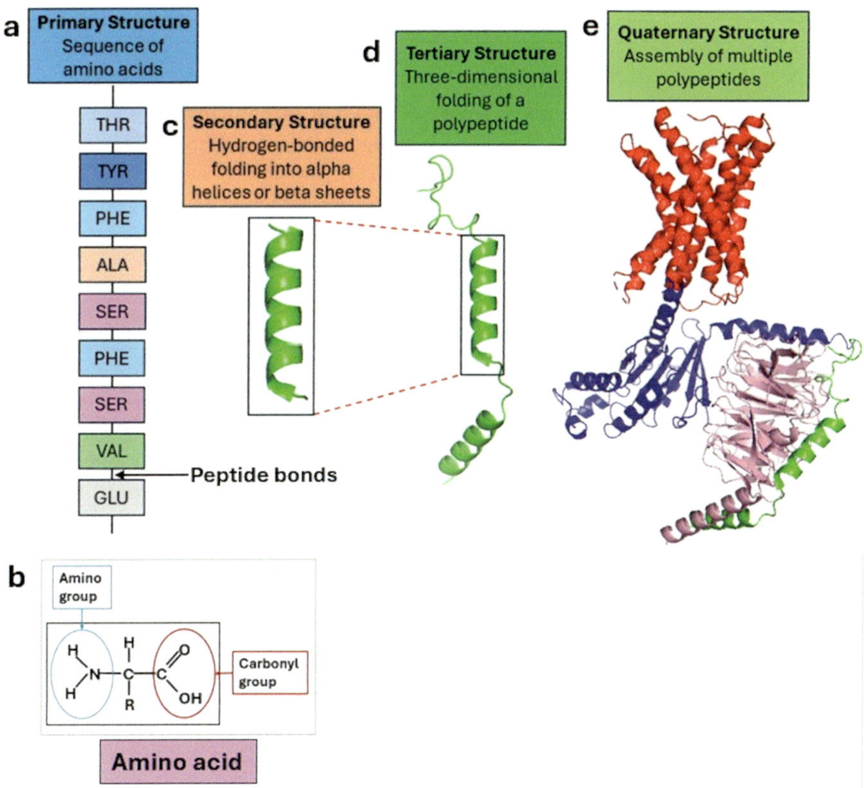

FIGURE 3.2 An image of the primary, secondary, tertiary, and quaternary structure of a protein. (Adapted from Xie et al. [74]). (a) Primary structure with sequence of amino acids; (b) amino acid structure; (c) secondary structure with folding into alpha helices and beta sheets; (d) tertiary structure with folding of polypeptide; (e) quaternary structure with assembly of multi-peptides.

a subset of amide bonds, which play a central role in protein structure and MIR analysis.

2. ***Secondary structure***: Stabilized by hydrogen bonds, forming α-helices and β-pleated sheets (Figure 3.2c).
3. ***Tertiary structure***: Three-dimensional folding due to interactions between side chains (R-groups) (Figure 3.2d).
4. ***Quaternary structure***: Assembly of multiple polypeptide subunits (e.g., hemoglobin) (Figure 3.2e).

3.3.4.2.1.4 Amide Bonds and Their MIR Signatures Amide bonds contain a carbonyl group (C = O) with a carbon (C) atom on one side and a nitrogen (N) atom on the other (Figure 3.3) [69]. The carbonyl group and an N both possess non-bonding electron pairs. These electrons conjugate along the C–N bond, creating a partial double-bond character and making the bond rigid. This rigidity contributes to the formation of secondary protein structures: α-helices and β-sheets. Conjugation also shifts the C = O stretch frequency to a lower wavenumber (1680–1630 cm^{-1}), compared to that in ketones or esters [69].

There are three types of amides [69]:

1. *Primary amides*: N bonded to two H atoms (Figure 3.3).
2. *Secondary amides*: N bonded to one H and one carbon (Figure 3.3).
3. *Tertiary amides*: N bonded to two carbons, with no N–H bond (Figure 3.3).

Proteins primarily contain secondary amide bonds, which exist in cis or trans configurations; trans being more common and typically analyzed in MIR spectroscopy [69].

Secondary amide bonds are as follows [69]:

1. *N–H stretch*: One band observed in the 3370–3170 cm^{-1} range. In contrast, primary amides show two bands due to two N–H bonds. The N–H stretch band is influenced by both the concentration of N–H groups and their bonding state—whether they are free or involved in hydrogen bonding. In other words, the position and shape of the MIR absorption band vary depending on the N–H concentration and whether the N–H bonds are free or bonded.
2. *C = O stretch (Amide I Band)*: Seen between 1680 and 1630 cm^{-1}.

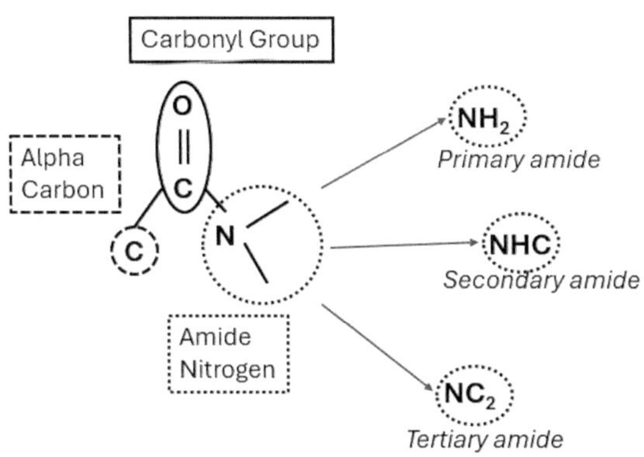

FIGURE 3.3 Amide structure and types of amide bonds.

3. *N–H in-plane bend (Amide II band)*: Intense band between 1570 and 1515 cm^{-1} (solid state) or slightly lower in solution.
4. *C–N stretch (Amide II band)*: Appears in the 1310–1230 cm^{-1} range, lower than that in primary amides due to interaction with N–H bending.
5. *N–H out-of-plane bend*: A broad band observed between 750 and 680 cm^{-1}, overlapping with primary amide bands.
6. Generally, the presence of carbonyl (C=O) stretches along with the N–H bend band confirms the presence of secondary amide structure.

Key MIR absorption bands for secondary amides are as follows [28]:

1. Amide I Band is seen due to C = O stretch in the region of 1650 cm^{-1} for a solid state of the sample and between 1715 and 167 cm^{-1} for a solution form.
2. Amide II Band is formed due to the mixed vibrations of N–H in-plane bending and C–N stretching. The physical form of the sample has a strong correlation with the band location. The band is seen in the 1570–1510 cm^{-1} range in solid forms and in the 1550–1500 cm^{-1} range for the solutions.
3. *Amide III band*: 1310–1200 cm^{-1}, due to combined C–N stretching and N–H bending.
4. *Amide IV band*: Near 620 cm^{-1}, related to O = C–N bending.
5. *Amide V band*: Around 700 cm^{-1}, attributed to N-H out-of-plane bending.

Among these, Amide I and Amide II bands are most frequently analyzed for protein structure elucidation due to their sensitivity to secondary structural elements.

3.3.4.2.1.5 Mid-Infrared (MIR) Spectroscopy in Biologics Manufacturing MIR spectroscopy, particularly ATR–FTIR, is gaining recognition for its utility in biologics manufacturing. The following studies highlight its growing application in monitoring CQAs and optimizing purification processes.

3.3.4.2.1.6 Case Study: In-Column ATR–FTIR for Monitoring Mab Purification and Cleaning-in-Place (CIP) Protein A affinity chromatography is widely used for the downstream purification of monoclonal antibodies. However, the resin's binding capacity diminishes with repeated use, primarily due to the accumulation of non-eluting proteins. The reduction in the binding capacity leads to resin replacement, increasing production costs. Cleaning-in-place (CIP) protocols are implemented to restore binding efficiency, but their effectiveness can vary.

Byrne et al. [29] proposed an in-column ATR–FTIR approach to monitor both purification and CIP processes. By integrating an ATR–FTIR sensor into an affinity chromatography column, the authors enabled real-time, in situ analysis of Protein A resin fouling during CIP cycles.

In this setup, a plunger was used to position stationary resin beads over the ATR detector. MIR spectra were collected at various stages (Figure 3.4a):

1. After mAb binding
2. After elution with low pH (pH 3.0)
3. After CIP using 50-mM NaOH

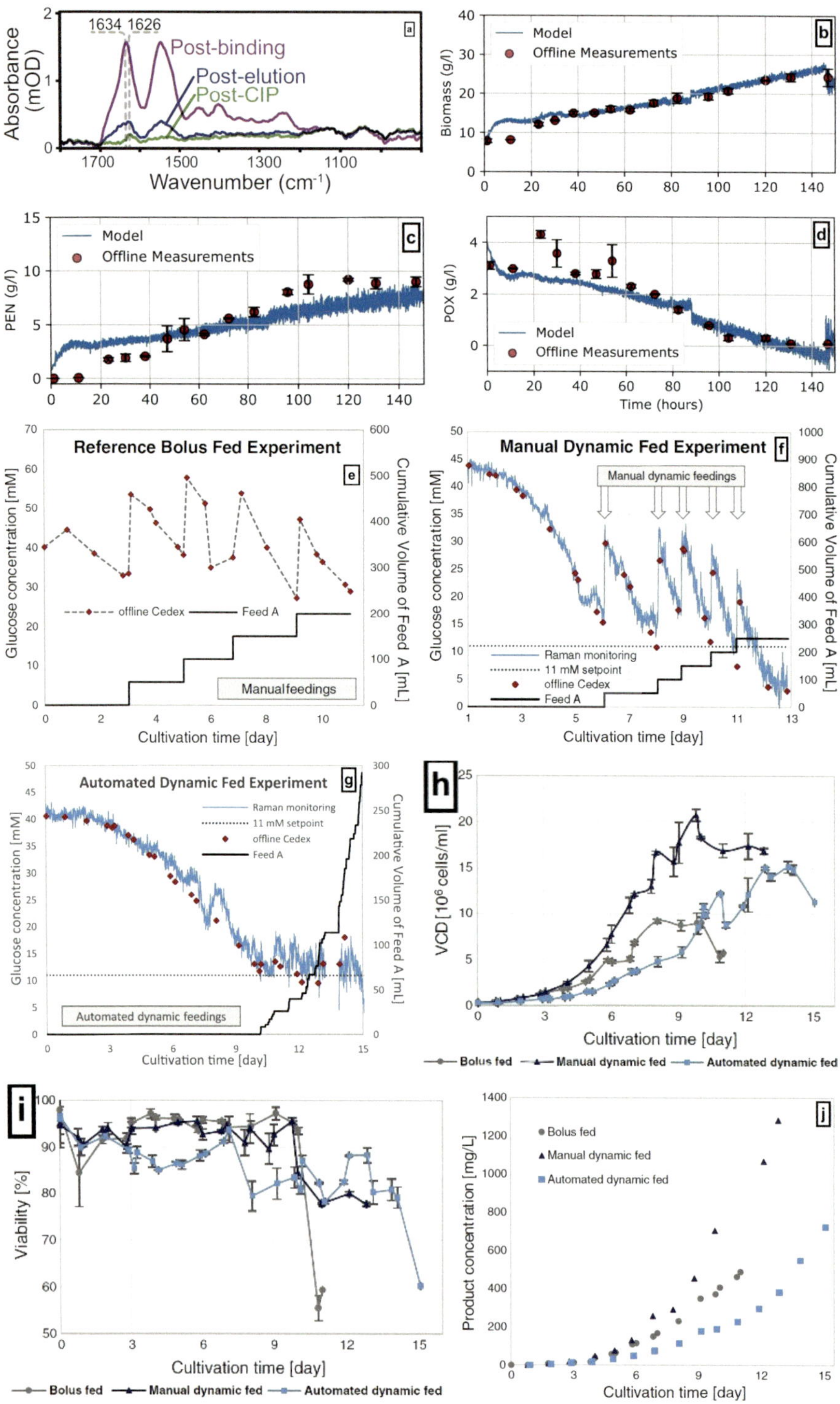

FIGURE 3.4 (*Continued*)

The resulting spectra featured distinct Amide I and II peaks. The Amide I band (1634–1626 cm^{-1}) in the post-binding stage (purple line) reflected the α-helical structure of the pure monoclonal antibody. Following elution (blue line) and CIP (green line), the Amide I peak shifted to lower wavenumbers, indicating the presence of β-sheet structures, characteristics of the residual contaminant proteins.

This shift confirmed that non-eluting β-structured proteins remain bound to the resin even after CIP, leading to decreased binding efficiency. The study concluded that current CIP protocols are insufficient for completely removing these contaminants.

The authors recommended ATR–FTIR as a non-invasive, label-free technique to:

1. Assess Protein A resin integrity and binding capacity
2. Optimize CIP strategies
3. Monitor contaminant build-up for improved process control and quality assurance

This study demonstrates that MIR spectroscopy, particularly FTIR and ATR–FTIR, offers significant advantages for biologics manufacturing. In addition to providing information about the secondary structure of the proteins, FTIR can quantify protein content based on Beer–Lambert's law [69]. From profiling glycosylation patterns, identifying impurities, protein aggregates, and ligand binding to monitoring purification processes in real time, MIR methods provide rapid, sensitive, and minimally invasive alternatives to conventional analytical tools. While they may not replace high-resolution methods like LC-MS for drug development, they serve as powerful tools for real-time quality control and process optimization in biopharmaceutical production.

3.3.4.2.1.7 Near-Infrared Spectroscopy (NIR) The NIR region spans 750–2500 nm (14,000–4000 cm^{-1}). In this range, molecular absorbance arises primarily from overtone transitions ($v = 0 \rightarrow 2,3,4$) and combination bands [69]. The absorbance intensity decreases significantly with overtone order—for instance, the $v = 0 \rightarrow 2$

FIGURE 3.4 (*Continued*) Case studies for spectroscopy-based tools: (a) *Mid-infrared spectroscopy*: the MIR spectrum showcases Amide I peaks after mAb binding (Post-binding; purple peaks), after elution (Post-elution; blue peaks), and after CIP (Post-CIP; green peaks). While purple peaks are indicative of the α-helical structure of pure mAb, blue and green peaks are shifting to lower wavenumbers (compared to the purple peaks) indicating the presence of β-sheet structures post-elution and post-CIP; (b–d) *Near infrared spectroscopy*: Comparing the off-line values (Red circles) of biomass (b), Penicillin (PEN) (c), and Phenoxyacetate (POX) (d) to the PLS regression model predicted values (Blue lines). There is an overestimation of biomass and PEN for the initial 20 h but the POX values are very well aligned; (e–g) *Raman spectroscopy*: comparison of three feeding strategies shows that the automated dynamic feeding (g) was able to maintain the glucose concentration closest to the optimal concentration of 11 mM followed by manual dynamic feeding (f). The bolus-fed cultures (e) always showed the glucose concentration of >30 mM; (h) VCD significantly improved with both the dynamic feed strategies than with the conventional bolus-fed method; (i) Cell viability was between 75 and 100% for all the feeding methods; (j) Improved product titer is seen with the dynamic feeding strategies than the conventional bolus feed method. ((a) Images from Byrne et al. [29], (b–d) images from Brandstetter et al. [75], and (e–g) images from Hirsch et al. [76]).

transition is approximately 10× weaker, and $v = 0 \rightarrow 3$ is ~100× weaker, than the fundamental $v = 0 \rightarrow 1$ transition [69].

NIR spectroscopy is particularly effective for probing X–H bonds (C–H, N–H, O–H), which produce the most informative absorptions in this region [68, 69]. Some lower-frequency modes, such as C = O and C = C stretches, may also be accessible [68]. As in MIR spectroscopy, spectra can be collected in absorbance or transmittance mode; however, for quantitative applications using multivariate data analysis, absorbance mode is used.

The inherently low absorbance in NIR enables deeper sample penetration, making it suitable for analyzing complex matrices [68, 77]. Combined with its ability to pass through glass, NIR has enabled the development of non-immersive probes that monitor biomass and protein production directly through glass bioreactor walls [75]. Low absorbance also facilitates analysis of highly scattering or absorbing systems (e.g., cell culture media) and allows for direct, preparation-free measurements [68, 78, 79].

3.3.4.2.1.7.1 NIR for Process Monitoring In biomanufacturing, monitoring critical metabolites such as glucose, ammonia, glutamate, and glutamine is essential [79]. NIR spectroscopy can track these metabolites via their X–H overtone and combination band signatures [79]. This enables online monitoring of substrates, metabolites, and biomass [78].

3.3.4.2.1.7.2 Challenges in Bioprocess Monitoring with NIR Two major challenges arise when using NIR for process monitoring:

1. *Water interference*: Strong water absorption bands near 5000 and 7000 cm^{-1} can obscure weaker signals from metabolites such as glucose. These regions can be excluded manually or via multivariate data analysis [80].
2. *Model development*: Building accurate Partial Least Squares (PLS) models is complex, and poorly fitted models can compromise monitoring accuracy [75, 79, 80].

3.3.4.2.1.7.3 NIR in Biologics Manufacturing: A Case Study on Biomass Monitoring Brandstetter et al. [75] demonstrated the use of non-immersive NIR sensors to measure biomass, penicillin (product), and phenoxyacetic acid (precursor) concentrations in *Penicillium chrysogenum* fed-batch fermentation (Figure 3.4b–d). Two spectral regions of interest (1350–1650 nm and 1550–1950 nm) were analyzed using PLS regression. Predicted values (PLS regression) aligned closely with off-line reference measurements, although overestimation (for biomass and PEN) occurred during the initial 20 h, underscoring the importance of precise model calibration.

3.3.4.2.1.8 Raman Spectroscopy While MIR and NIR spectroscopies are based on the absorption of electromagnetic radiation, Raman spectroscopy operates on the principle of light scattering. Specifically, it exploits the Raman effect, which states that the frequency of scattered radiation differs from that of the incident monochromatic radiation [70]. Raman measurements are typically conducted in the 4000–400 cm^{-1} wavenumber range, similar to MIR spectroscopy, but differ because the analytical information is gained through the scattering of radiation rather than the absorbance [70].

When monochromatic radiation interacts with molecules, most of the scattered light retains the same frequency as the incident beam; this is known as Rayleigh scattering, which contains no chemical information [70, 81]. A small fraction of the scattered light has a different frequency due to inelastic scattering, which provides molecular vibrational information [70, 81]. If the scattered light has a lower frequency than the incident radiation, the process is called Stokes Raman scattering, producing Stokes lines in the spectrum [70, 81]. Conversely, scattered light with a higher frequency is called anti-Stokes Raman scattering, producing anti-Stokes lines [70, 81]. Stokes lines are typically more intense and are most often used in conventional Raman spectroscopy [70, 81].

The difference in frequency between the incident and scattered light is known as the Raman shift, which directly selects the vibrational energy levels of the molecule responsible for the scattering [70, 81].

a. *Advantages*—Raman spectroscopy is simpler to interpret than infrared spectroscopy because overtone and combination bands are absent [82]. Furthermore, it is unaffected by water interference, making it advantageous for aqueous systems. Raman spectroscopy can analyze samples in various states—liquid, solid, gas, or gel—using both contact and non-contact probes [82].

As with infrared methods, Raman spectroscopy data for bioprocess monitoring are typically analyzed using multivariate data analysis. Principal Component Analysis (PCA) followed by PLS regression is commonly applied to develop predictive models and enable automated process control.

3.3.4.2.1.9 Case Study: Glucose Monitoring and Automated Feeding In mAb production using mammalian cell cultures, glycosylation is usually a CQA, and glucose concentration is a CPP influencing this CQA. As described earlier, various analytical techniques have been used to monitor glucose and related metabolites to ensure optimal product quality.

Hirsch et al. [76] implemented in-line Raman spectroscopy to maintain glucose concentrations around 11 mM and compared three feeding strategies in Chinese Hamster Ovary (CHO) cell cultures:

- Reference bolus feeding (manual, based on off-line measurements)
- Manual dynamic feeding (manual, based on in-line measurements, predetermined feed volumes)
- Automated dynamic feeding (automated, based on real-time in-line glucose measurements)

The target concentration of 11 mM was chosen to prevent glucose depletion, enhance cell viability and productivity, and limit lactate accumulation.

Results

- *Glucose concentration control (Figure 3.4e–g)*:
  - Reference bolus feeding maintained glucose above 30 mM.
  - Manual dynamic feeding maintained glucose between 10 and 30 mM.

    - Automated dynamic feeding held glucose within the optimal 10–15 mM range.
- *VCD (Figure 3.4h and 3.4i)*:
    - Both dynamic strategies (manual and automated) significantly improved VCD compared with bolus feeding.
    - Cell viability remained between 75 and 100% in all fed batches.
    - A temporary drop in the automated dynamic run was due to an automation default on day 14.
- *Product titer (Figure 3.4j)*:
    - Reference bolus: 0.49 g/L
    - Manual dynamic: 1.28 g/L
    - Automated dynamic: 0.72 g/L

These results demonstrate that dynamic feeding informed by real-time Raman measurements can improve cell culture performance and product yield compared with traditional bolus feeding, although feed strategy optimization is required to maximize benefits.

3.3.5 Other Tools for PAT

While physico-chemical, biomolecular, dielectric, and spectroscopy-based tools form the core of modern PAT, there is a growing class of complementary technologies that address specific analytical needs not easily met by conventional methods. These "other" PAT tools are diverse but share three defining characteristics:

a. Specialized measurement principles that target unique product or process attributes.
b. Real-time or near-real-time capability for monitoring within the manufacturing environment.
c. Integration potential with multi-sensor strategies to create a more comprehensive process understanding.

This category includes particle imaging systems, mass spectrometry (MS), calorimetry, FBRM variants, and hybrid sensors that combine multiple analytical modalities. Their role is often critical in scale-up, formulation development, and complex bioprocesses such as vaccines, where multiple CQAs must be controlled simultaneously.

3.3.5.1 Particle Imaging and Morphological Tools

Particle Vision Monitoring (PVM) employs in-situ high-resolution imaging to visualize particles or aggregates in real time [83, 84]. In vaccine formulation, PVM can detect morphological changes in adjuvant particles during antigen adsorption, providing early warning of aggregation events. When integrated with FBRM, it offers both size distribution data and morphological context, enabling more robust process control [85].

Micro-Flow Imaging (MFI) is another valuable technique for detecting and counting subvisible particles in biopharmaceutical formulations [86]. In therapeutic protein manufacturing, MFI can detect proteinaceous aggregates and silicone oil droplets in pre-filled syringes, both of which are critical CQAs for parenteral products [87, 88].

3.3.5.2 Particle Vision Monitoring (PVM)

PVM uses a high-resolution in-situ camera with controlled illumination to capture real-time images of particles or aggregates in a process stream. Applications in PAT are: (a) monitoring morphological changes in vaccine adjuvants during antigen adsorption, (b) tracking crystallization dynamics during API manufacturing, and (c) detecting oil droplet size and distribution in emulsion-based formulations. It provides direct visual confirmation of particle shape, size, and texture. Image interpretation can be subjective and requires image analysis software for quantitative use. Combined PVM + FBRM systems allow simultaneous measurement of PSD (via chord length) and morphology. In vaccine formulation, this helps differentiate between normal adsorption profiles and aggregation events that may compromise potency [89, 90].

3.3.5.3 Mass Spectrometry (MS)

MS, while traditionally off-line, is increasingly being adapted for on-line or at-line monitoring in Bioprocessing. For example, in microbial fermentation, MS can monitor volatile metabolites in the headspace, providing insights into metabolic shifts. In chemical synthesis of small molecule adjuvants, on-line MS can detect residual solvents or impurities in real time [3, 91]. It ionizes analytes and separates ions based on mass-to-charge ratio (m/z) for highly specific chemical identification and quantification. While MS has traditionally been an off-line laboratory tool, on-line MS configurations are emerging, especially in continuous manufacturing. Applications include: (a) monitoring volatile components in headspace gases during fermentation, (b) tracking process impurities such as solvents, degradation products, or leachables, and (c) real-time reaction monitoring in chemical API synthesis. It is extremely specific, detects trace levels of analytes, and has high complexity, maintenance needs, and sensitivity to fouling.

3.3.5.4 Calorimetry

Calorimetry, particularly reaction calorimetry, can be used to monitor metabolic activity in fermentation via heat output or to ensure safety during exothermic chemical steps in adjuvant production. While not molecule-specific, calorimetry provides a non-invasive way to track overall process kinetics and measures heat flow associated with chemical or biochemical reactions to infer reaction rate and extent. Various applications in PAT include (a) detecting metabolic activity in fermentation via heat release profiles and (b) monitoring exothermic or endothermic steps in chemical synthesis to ensure safety and consistency. They are non-invasive and work regardless of optical properties. Less specifically, changes in heat flow require interpretation alongside other measurements. In continuous Bioprocessing, calorimetric data can feed into feed-rate control algorithms, maintaining metabolic rates within optimal ranges.

3.3.5.5 Hybrid and Multi-Modal Sensors

Hybrid sensors that combine multiple measurement modalities—such as FBRM with Raman spectroscopy or dielectric probes with NIR spectroscopy—offer a powerful approach to holistic process monitoring. These systems reduce the number of bioreactor ports needed while providing complementary datasets for multivariate analysis. Devices that combine multiple measurement principles into a single in-situ probe reduce the number of process ports needed and provide richer data for multivariate analysis. Examples of modes of hybrid and multi-modal sensors include the following:

- ***FBRM + Raman***: Simultaneous particle size and chemical fingerprinting.
- ***Turbidity + conductivity***: Combined monitoring of filtration breakthrough and ionic strength changes.
- ***Dielectric + NIR***: Linking biomass growth with nutrient consumption rates.

The above description of the multiple tools associated with PAT is summarized as principle advantages, and disadvantages in Table 3.1.

TABLE 3.1
Summary of the Various Tools Introduced Here, along with Their Principle of Operation

Technique	Principle	Advantages	Limitations	Key Bioprocessing Area(s)
Temperature sensors (thermocouple/ RTD)	Electrical resistance/ thermo-EMF vs. T	Robust, accurate, continuous	Probe drift; sterilization/ placement constraints	Upstream (bioreactors); downstream (chromatography temp control); fill–finish
pH probes	Glass electrode potential ∝ $[H^+]$	Direct CQA/CPP linkage; fast	Fouling, calibration, CO_2 effects	Upstream (fermentation/cell culture); downstream (chromatography buffers)
Dissolved oxygen (DO)	Electrochemical/ optical quenching by O_2	Real-time oxygen transfer insight	Membrane fouling; calibration	Upstream (aeration control); downstream (folding steps)
Conductivity	Solution ionic strength via σ	Simple, rugged, online	Non-specific; temp-dependent	Downstream (column control, buffer dilution); upstream media checks
Osmometry	Colligative property vs. solute	Viability indicator	Off/at-line mostly; sample needed	Upstream (mammalian culture)
Viscosity	Shear-dependent torque/frequency response	Early deviation detection (UF/ DF, mixing)	Non-specific; shear-history effects	Downstream (UF/DF); formulation
Turbidity/Optical density	Light attenuation/ scattering vs. particles	Inexpensive, easy integration	Non-specific; bubbles/foam interference	Upstream (biomass, feed control); downstream (filter breakthrough)

(Continued)

TABLE 3.1 (*Continued*)
Summary of the Various Tools Introduced Here, along with Their Principle of Operation

Technique	Principle	Advantages	Limitations	Key Bioprocessing Area(s)
Focused beam reflectance measurement (FBRM)	Backscattered laser chord-length counts	Real time in opaque slurries; nucleation/ growth tracking	Chord length ≠ true size; morphology bias	Upstream (adjuvant, crystallization); downstream (protein crystallization)
Laser diffraction (LD)	Angular scattering → size distribution	Established models; broad size range	Usually off-line; prep bias	Upstream (adjuvant QC); formulation
In-situ microscopy/ EasyViewer (PVM family)	In-situ imaging of particles/ aggregates	Morphology + visual confirmation	Image analysis burden; fouling	Upstream (adjuvant–antigen); downstream (crystallization)
Soft sensors/ chemometrics	Models map sensor signals → latent variables	Estimates hard-to-measure CQAs; predictive	Model maintenance; data drift	Cross-stage (advanced control, digital twins)
ELISA/ flow-ELISA	Antigen–antibody binding → optical readout	High specificity; at/online variants	Reagents, fouling; often off-line	Upstream (product titer), downstream (HCP), fill–finish (potency)
DLS	Intensity fluctuations → hydrodynamic size	Rapid sub-μm sizing; aggregation sensing	Dilution/ polydispersity artifacts	Upstream (antigen–adjuvant size); formulation (LNPs, protein stability)
MALS (incl. SEC-MALS)	Multi-angle scattering → absolute Mw/Rg	Calibration-free Mw; aggregate detection	Needs optical clarity; dust-sensitive	Downstream (polishing, purity/ aggregate control)
BLI/SPR	Label-free reactive/ optical interference at the surface	Real-time kinetics (ka/kd/KD)	At-line; surface fouling in crude	Upstream PD (binding verification), formulation (stability, comparability)
Nucleic acid assays (UV260, PicoGreen, CE)	Spectroscopy/ fluorescence; electrophoretic integrity	Simple, sensitive (dsDNA); integrity profiling	Clarification/ standards; at/ off-line	Upstream/downstream (AAV, plasmid, mRNA); formulation (LNP concentration)
Dielectric spectroscopy (capacitance)	Complex permittivity vs. freq; viable cell polarization	Non-invasive VCD; continuous; feedback control	Salt/temperature effects; calibration	Upstream (cell growth, apoptosis, harvest timing)
Mid-IR/FTIR/ ATR–FTIR	Fundamental vibrations (absorbance)	Structural "fingerprints"; inline ATR; MAM potential	Water interference; probe integration	Upstream (protein structure); downstream (column fouling/CIP); formulation
Near-IR (NIR)	Overtones/ combination bands	Penetrates turbid media; through glass; multivariate	Strong water bands; model building	Upstream (substrates, biomass); downstream (liquid streams)

(*Continued*)

TABLE 3.1 (*Continued*)
Summary of the Various Tools Introduced Here, along with Their Principle of Operation

Technique	Principle	Advantages	Limitations	Key Bioprocessing Area(s)
Raman	Inelastic scattering (Raman shift)	Water-insensitive; inline; simpler banding	Fluorescence background; SNR at low conc.	Upstream (glucose control, automated feeding); cross-stage PAT
Mass spectrometry (process)	m/z-resolved ion detection	High specificity; volatiles, impurities	Complexity; fouling; maintenance	Upstream (off-gas), downstream (impurities); continuous synthesis
Calorimetry (reaction/heat flow)	Heat release/ absorption vs. rate	Non-specific, universal kinetic signal	Interpretation needs context; sensitivity	Upstream (metabolic activity); continuous synthesis safety
Hybrid probes (e.g., FBRM+Raman; Dielectric+NIR)	Co-located multi-modal sensing	Complementary data; fewer ports	Cross-interference; co-calibration	Cross-stage (holistic control, multivariate models)

3.4 APPROACHES FOR PAT IMPLEMENTATION

As described earlier, biologic manufacturing typically involves three key stages: upstream processing, downstream processing, and final formulation [67, 92, 93]. To enhance quality standards within the pharmaceutical industry, the U.S. Food and Drug Administration (FDA) introduced PAT [1, 2]. PAT is a framework outlined to facilitate the design, analysis, and control of biomanufacturing processes. It integrates a range of analytical techniques implemented at critical control points (CCPs) to monitor CPPs, which are identified based on CQAs [1, 2].

Unlike traditional manufacturing, where analyses are typically conducted on raw materials, intermediates, or final products, PAT emphasizes real-time monitoring throughout the manufacturing process [1, 2, 72]. In QbD into biomanufacturing [1, 2], process monitoring strategies—eventually leading to process control—can be implemented at different levels: off-line, at-line, on-line, or in-line [71].

a. **Off-line monitoring** involves withdrawing samples from the bioreactor and transporting them to an external analytical facility for testing. This approach often results in delays, limiting its usefulness for real-time process adjustments. Frequent sampling also requires human intervention, which increases the risk of contamination, while the sampling frequency is typically lower compared to other monitoring methods.
b. **At-line monitoring** also requires sampling, but analyses are conducted at the same facility, reducing the risk of contamination during transport.

Results are generally available within an hour, allowing process adjustments to be implemented within a few hours.

c. **On-line monitoring** is directly integrated with the bioreactor, where samples are automatically diverted for analysis without human intervention. This minimizes contamination risks and provides results within 5–10 min, enabling relatively rapid adjustments to the biomanufacturing process.
d. **In-line monitoring** places sensors and analytical tools directly inside the bioreactor, enabling fully automated and continuous data collection. This approach eliminates human error and allows for real-time process adjustments. However, it relies heavily on advanced modeling tools for data interpretation and process control.

PAT implementation with QbD allows for in-line implementation of analytical tools—although at-line and on-line methods are also used—offering enhanced process understanding and control. However, full realization of PAT's automation potential is best achieved through in-line deployment.

3.5 CONCLUSION

PAT provides a coherent, science-based pathway to real-time understanding and control of biologics manufacturing. This chapter outlined the PAT workflow linking target product profile, CPPs, and CQAs to monitoring strategies, performance criteria, tool selection, and model validation. Complementary approaches—monitoring of cell state, process streams, and culture environment—are enabled by physico-chemical sensors or core PAT parameters. Product-specific PAT measurements are enabled by biomolecular assays, dielectric spectroscopy, and vibrational spectroscopics, augmented by particle imaging, MS, calorimetry, and hybrid probes. Case studies across vaccines, product crystallization, chromatography, and fermentation showed how in-line and at-line analytics illuminate mechanisms, quantify kinetics, and detect failures early. Chemometric soft sensors integrate multivariate signals to estimate hard-to-measure attributes and to drive feedback control. Implementation across off-line to in-line modalities under QbD demonstrates measurable gains in robustness, yield, and compliance. Establishing traceable links from CPPs to CQAs creates the data foundation for digital twins, continuous verification, and ultimately safe, autonomous process control in next-generation, intensified biomanufacturing.

REFERENCES

1. Chopda, V., et al., Recent advances in integrated process analytical techniques, modeling, and control strategies to enable continuous biomanufacturing of monoclonal antibodies. Journal of Chemical Technology & Biotechnology, 2021. **97**(9): p. 2317–2335.
2. Chanda, A., et al., Industry perspectives on process analytical technology: Tools and applications in API development. Organic Process Research & Development, 2014. **19**(1): p. 63–83.
3. Randek, J. and Mandenius C.F., *On-line soft sensing in upstream bioprocessing.* Critical Reviews in Biotechnology, 2018. **38**(1): p. 106–121.

4. Gustavsson, R. and Mandenius C.F., Soft sensor control of metabolic fluxes in a recombinant *Escherichia coli* fed-batch cultivation producing green fluorescence protein. Bioprocess and Biosystems Engineering, 2013. **36**(10): p. 1375–84.
5. Rusli, N.I., et al., *Multi-sensor chip for monitoring key parameters in bioprocesses.* IEEE Sensors Journal, 2021. **21**(18): p. 19675–19683.
6. Todtenberg, N., et al., *Autonomous sensor capsule for usage in bioreactors.* IEEE Sensors Journal, 2015. **15**(7): p. 4093–4102.
7. Rathore, A.S., R. Bhambure, and V. Ghare, *Process analytical technology (PAT) for biopharmaceutical products.* Analytical and Bioanalytical Chemistry, 2010. **398**(1): p. 137–154.
8. Simon, L.L., et al., *Assessment of recent process analytical technology (PAT) trends:* A *multiauthor review.* Organic Process Research & Development, 2015. **19**(1): p. 3–62.
9. Workman, Jr, J., et al., *Process analytical chemistry.* Analytical Chemistry, 2011. **83**(12): p. 4557–4578.
10. Wasalathanthri, D.P., et al., Technology outlook for real-time quality attribute and process parameter monitoring in biopharmaceutical development—A review. Biotechnology and Bioengineering, 2020. **117**(10): p. 3182–3198.
11. Aifuwa, I., *PAT strategies and applications for cell therapy processing.* Current Opinion in Biomedical Engineering, 2022. **24**: p. 100405.
12. Schmidt, A., et al., Process automation and control strategy by quality-by-design in total continuous mRNA manufacturing platforms. Processes, 2022. **10**(9): p. 1783.
13. Zobel-Roos, S., et al., Accelerating biologics manufacturing by modeling or: Is approval under the QbD and PAT approaches demanded by authorities acceptable without a digital-twin? Processes, 2019. **7**(2): p. 94.
14. Pomerantsev, A.L. and Rodionova O.Y., *Process analytical technology: A critical view of the chemometricians.* Journal of Chemometrics, 2012. **26**(6): p. 299–310.
15. Seenivasan R. et al., A systematic review: Exploration of process analytical technology techniques (PAT) and their multifaceted advantages in industrial processes. International Journal of Applied Pharmaceutics, 2024. 16(2): p. 44–51.
16. Rolinger, L., Rüdt M., and Hubbuch J., A critical review of recent trends, and a future perspective of optical spectroscopy as PAT in biopharmaceutical downstream processing. Analytical and Bioanalytical Chemistry, 2020. **412**(9): p. 2047–2064.
17. Thakur, G., S. Thori, and Rathore A.S. Implementing PAT for single-pass tangential flow ultrafiltration for continuous manufacturing of monoclonal antibodies. Journal of Membrane Science, 2020. **613**: p. 118492.
18. Knop, K. and Kleinebudde P., *PAT-tools for process control in pharmaceutical film coating applications.* International Journal of Pharmaceutics, 2013. **457**(2): p. 527–536.
19. Whitford, W. and Julien C., *Analytical technology and PAT.* BioProcess International, 2007. **5**: p. S32–S45.
20. Vetter, F.L., Zobel-Roos S., and Strube J., PAT for continuous chromatography integrated into continuous manufacturing of biologics towards autonomous operation. Processes, 2021. **9**(3): p. 472.
21. Mitra, S. and Murthy G.S., *Bioreactor control systems in the biopharmaceutical industry:* A *critical perspective.* Systems Microbiology and Biomanufacturing, 2022. **2**(1): p. 91–112.
22. Urthaler, J., et al., *Industrial manufacturing of plasmid-DNA products for gene vaccination and therapy, in Gene vaccines.* 2011, Springer. p. 311–330.
23. Clayton, K.N., et al., DNA microviscosity characterization with particle diffusometry for downstream DNA detection applications. Analytical Chemistry, 2017. **89**(24): p. 13334–13341.
24. Qwist, P.K., et al., *On-line rheological characterization of semi-solid formulations.* European Journal of Pharmaceutical Sciences, 2019. **128**: p. 36–42.

25. Ebert, F.V., et al., Characterization of a noninvasive on-line turbidity sensor in shake flasks for biomass measurements. Biochemical Engineering Journal, 2018. **132**: p. 20–28.
26. Pizarro, S.A., et al., Biomanufacturing process analytical technology (PAT) application for downstream processing: Using dissolved oxygen as an indicator of product quality for a protein refolding reaction. Biotechnology and Bioengineering, 2009. **104**(2): p. 340–351.
27. Tok, A.T., et al., Monitoring granulation rate processes using three PAT tools in a pilot-scale fluidized bed. AAPS PharmSciTech, 2008. **9**(4): p. 1083–1091.
28. Parker, F.S., *Amides, in infrared spectroscopy-in niochemistry, biology, and medicine*. 1971, Plenum Press. p. 165.
29. Boulet-Audet, M., Kazarian S.G., and Byrne B., In-column ATR–FTIR spectroscopy to monitor affinity chromatography purification of monoclonal antibodies. Scientific Reports, 2016. **6**: p. 30526.
30. Walla, B., et al., Recent advances in the monitoring of protein crystallization processes in downstream processing. Crystals, 2023. **13**(5): p. 773.
31. Hu, C., et al., Continuous reactive crystallization of an API in PFR–CSTR cascade with in-line PATs. Reaction Chemistry & Engineering, 2020. **5**(10): p. 1950–1962.
32. Leite, L.C. and A. Bernardo, A new alternative for monitoring the crystallization process through artificial neural networks and FBRM signal. 2023.
33. Ruf A., Worlitschek J., and Mazzotti M., *Modeling and experimental analysis of PSD measurements through FBRM*. Particle & Particle Systems Characterization, 2000. **17**(4): p. 167–179.
34. Kowal, P., et al., *Application of PAT probes in* aluminum phosphate adjuvant manufacturing. Pharmaceutical Research, 2024. **41**(2): p. 375–385.
35. Shekunov, B.Y., et al., *Particle size analysis in pharmaceutics: Principles, methods and applications*. Pharmaceutical Research, 2007. **24**(2): p. 203–227.
36. de Boer, G.B., et al., *Laser diffraction spectrometry: Fraunhofer diffraction versus Mie scattering*. Particle & Particle Systems Characterization, 1987. **4**(1–4): p. 14–19.
37. Silva, A.F., et al., Particle sizing measurements in pharmaceutical applications: Comparison of in-process methods versus off-line methods. European Journal of Pharmaceutics and Biopharmaceutics, 2013. **85**(3): p. 1006–1018.
38. Kitchener, B.G., J. Wainwright, and A.J. Parsons, *A review of the principles of turbidity measurement*. Progress in Physical Geography, 2017. **41**(5): p. 620–642.
39. Matos, T., et al., A review of methods and instruments to monitor turbidity and suspended sediment concentration. Journal of Water Process Engineering, 2024. **64**: p. 105624.
40. Bowers, D., et al., Light scattering by particles suspended in the sea: The role of particle size and density. Continental Shelf Research, 2009. **29**(14): p. 1748–1755.
41. Zanzotto, A., et al., *Membrane-aerated microbioreactor for high-throughput bioprocessing*. Biotechnology and Bioengineering, 2004. **87**(2): p. 243–254.
42. Adiga, R., et al., Point-of-care production of therapeutic proteins of good-manufacturing-practice quality. Nature Biomedical Engineering, 2018. **2**(9): p. 675–686.
43. Kekic, M., et al., Biosensing using antibody-modulated motility of actin filaments on myosin-coated surfaces. Biosensors and Bioelectronics, 2024. **246**: p. 115879.
44. Wang, W. and C.J. Roberts, *Aggregation of therapeutic proteins*. 2010, John Wiley & Sons.
45. Hu, C., et al., *Research progress on the quality control of mRNA vaccines*. Expert Review of Vaccines, 2024. **23**(1): p. 570–583.
46. Bhatnagar, A., et al., *Process development for biologics therapeutics, in approved: The life cycle of drug development*. 2025, Springer. p. 249–297.

47. Luo, D., et al., Lateral flow immunoassay for rapid and sensitive detection of dsRNA contaminants in in vitro-transcribed mRNA products. Molecular Therapy Nucleic Acids, 2023. **32**: p. 445–453.
48. Sandhu, R., et al., *Dynamic light scattering (DLS) technique, principle, theoretical considerations and applications.* Nano-Technological and Biochemical Techniques for Assessing the Quality and Safety of Milk and Milk Product, 2018, ICAR-NDRI. **8**: p. 135–7.
49. Yan, S., W. Gu, and Z.P. Xu, Re-considering how particle size and other properties of antigen–adjuvant complexes impact on the immune responses. Journal of Colloid and Interface Science, 2013. **395**: p. 1–10.
50. Yu, Z., J.C. Reid, and Y.-P. Yang, Utilizing dynamic light scattering as a process analytical technology for protein folding and aggregation monitoring in vaccine manufacturing. Journal of Pharmaceutical Sciences, 2013. **102**(12): p. 4284–4290.
51. São Pedro, M.N., et al., Process analytical technique (PAT) miniaturization for monoclonal antibody aggregate detection in continuous downstream processing. Journal of Chemical Technology & Biotechnology, 2022. **97**(9): p. 2347–2364.
52. Youssef, M., et al., Preliminary evaluation of formulations for stability of mRNA-LNPs through freeze-thaw stresses and long-term storage. 2025.
53. Khan, M.F.H., et al., *Freeze-drying of mRNA-LNPs vaccines: A review.* Vaccines, 2025. **13**(8): p. 853.
54. Falke, S. and C. Betzel, *Dynamic light scattering (DLS) principles, perspectives, applications to biological samples, in Radiation in bioanalysis: Spectroscopic techniques and theoretical methods.* 2019, Springer. p. 173–193.
55. Bhattacharjee, S., *DLS and zeta potential–what they are and what they are not?* Journal of Controlled Release, 2016. **235**: p. 337–351.
56. Meyer, R.A., et al., A scalable and robust cationic lipid/polymer hybrid nanoparticle platform for mRNA delivery. International Journal of Pharmaceutics, 2022. **611**: p. 121314.
57. Patel, B.A., et al. Multi-angle light scattering as a process analytical technology measuring real-time molecular weight for downstream process control, MAbs, 2018. 10(7): 945–950.
58. Matson, J.B., et al., Polymer characterization by size-exclusion chromatography with multi-angle light scattering (SEC-MALS): A tutorial review. Polymer Chemistry, 2024. **15**(3): p. 127–142.
59. Mattiasson, B., K. Teeparuksapun, and M. Hedström, Immunochemical binding assays for detection and quantification of trace impurities in biotechnological production. Trends in Biotechnology, 2010. **28**(1): p. 20–27.
60. Nilsson, M., et al., On-line monitoring of product concentration by flow-ELISA in an integrated fermentation and purification process. Journal of Fermentation and Bioengineering, 1994. **78**(5): p. 356–360.
61. Justice, C., et al., *Process control in cell culture technology using dielectric spectroscopy.* Biotechnology Advances, 2011. **29**(4): p. 391–401.
62. Heileman, K., J. Daoud, and M. Tabrizian, Dielectric spectroscopy as a viable biosensing tool for cell and tissue characterization and analysis. Biosensors and Bioelectronics, 2013. **49**: p. 348–359.
63. Asami, K., E. Gheorghiu, and T. Yonezawa, *Real-time monitoring of yeast cell division by dielectric spectroscopy.* Biophysical Journal, 1999. **76**(6): p. 3345–3348.
64. Yardley, J.E., et al., *On-line, real-time measurements of cellular biomass using dielectric spectroscopy.* Biotechnology and Genetic Engineering Reviews, 2000. **17**(1): p. 3–36.
65. Patel, P. and G.H. Markx, *Dielectric measurement of cell death.* Enzyme and Microbial Technology, 2008. **43**(7): p. 463–470.

66. Nikita, S., et al., *Advances in bioreactor control for production of biotherapeutic products*. Biotechnology and Bioengineering, 2023. **120**(5): p. 1189–1214.
67. Dhanuka, P., M.S.R. Wasalathanthri, M.W. Jay, M.C. Borys, J. Ding, and Z.J. Li, *Paving the way for real time process monitoring in biomanufacturing*, in *American pharmaceutical review*. 2020, CompareNetworks Inc. p. 54–57.
68. Bakeev, K.A., *Future trends in process analytical chemistry*. 2008, Analytical Technology. p. 424.
69. Smith, B., *Infrared spectral interpretation: A systematic approach*. 1999, CRC Press.
70. Bumbrah, G.S. and R.M. Sharma, Raman spectroscopy – Basic principle, instrumentation and selected applications for the characterization of drugs of abuse. Egyptian Journal of Forensic Sciences, 2016. **6**(3): p. 209–215.
71. Tiernan, H., B. Byrne, and S.G. Kazarian, *ATR–FTIR spectroscopy and spectroscopic imaging for the analysis of biopharmaceuticals*. Spectrochimica Acta Part A: Molecular and Biomolecular Spectroscopy, 2020. **241**: p. 118636.
72. Glassey, J., et al., *Process analytical technology (PAT) for biopharmaceuticals*. Biotechnology Journal, 2011. **6**(4): p. 369–77.
73. Hoehn, E.N.M.A.K., *Levels of organization*, in *Human anatomy and physiology*. 2019, Pearson. p. 28–38.
74. Chen, L., et al., AI-driven deep learning techniques in protein structure prediction. International Journal of Molecular Sciences, 2024. **25**(15).
75. Zimmerleiter, R., et al., Probeless non-invasive near-infrared spectroscopic bioprocess monitoring using microspectrometer technology. Analytical and Bioanalytical Chemistry, 2020. **412**(9): p. 2103–2109.
76. Domjan, J., et al., Raman-based dynamic feeding strategies using real-time glucose concentration monitoring system during adalimumab producing CHO cell cultivation. Biotechnology Progress, 2020. **36**(6): p. e3052.
77. Lundsberg-Nielsen L., Schlindwein W-S. and Berghaus A., Process Analytical Technology (PAT) Chapt 9. in *Pharmaceutical* quality by design – *A practical approach, Advances in pharmaceutical technology*, Edited by Schlindwein W-S. and Gibson M. 2018, Wiley
78. Scarff, M., et al., Near infrared spectroscopy for bioprocess monitoring and control: Current status and future trends. Critical Reviews in Biotechnology, 2006. **26**(1): p. 17–39.
79. Undey, C., et al., *Pat applied in biopharmaceutical process development and manufacturing: An enabling tool for quality-by-design*. 2011, CRC Press.
80. Kozma, B., A. Salgo, and S. Gergely, On-line glucose monitoring by near infrared spectroscopy during the scale up steps of mammalian cell cultivation process development. Bioprocess and Biosystems Engineering, 2019. **42**(6): p. 921–932.
81. Paul Rostron, S.G. and D. Gaber, *Raman spectroscopy, review*. International Journal of Engineering and Technical Research, 2016. **6**(1): p. 50–64.
82. Yang, N., et al., Raman spectroscopy applied to online monitoring of a bioreactor: Tackling the limit of detection. Spectrochimica Acta, Part A: Molecular and Biomolecular Spectroscopy, 2024. **304**: p. 123343.
83. Galata, D.L., et al., *Continuous blending monitored and feedback controlled by machine vision-based PAT tool*. Journal of Pharmaceutical and Biomedical Analysis, 2021. **196**: p. 113902.
84. Meszaros, L.A., et al., UV/VIS imaging-based PAT tool for drug particle size inspection in intact tablets supported by pattern recognition neural networks. International Journal of Pharmaceutics, 2022. **620**: p. 121773.
85. Arp, Z., et al., Optimization of a high shear wet granulation process using focused beam reflectance measurement and particle vision microscope technologies. Journal of Pharmaceutical Sciences, 2011. **100**(8): p. 3431–3440.

86. Fawaz, I., et al., *Micro-flow imaging multi-instrument evaluation for sub-visible particle detection*. European Journal of Pharmaceutics and Biopharmaceutics, 2023. **185**: p. 55–70.
87. Shah, M., et al., Evaluation of aggregate and silicone-oil counts in pre-filled siliconized syringes: An orthogonal study characterising the entire subvisible size range. International Journal of Pharmaceutics, 2017. **519**(1–2): p. 58–66.
88. Teska, B.M., et al., Aggregation and particle formation of therapeutic proteins in contact with a novel fluoropolymer surface versus siliconized surfaces: Effects of agitation in vials and in prefilled syringes. Journal of Pharmaceutical Sciences, 2016. **105**(7): p. 2053–2065.
89. Zhou, G., et al., *Process development and control with recent new FBRM, PVM, and IR*. Organic Process Research & Development, 2015. **19**(1): p. 227–235.
90. Jia, C.-Y., et al., Polymorphic transformation of pravastatin sodium monitored using combined online FBRM and PVM. Organic Process Research & Development, 2008. **12**(6): p. 1223–1228.
91. Zhao, L., et al., *Advances in process monitoring tools for cell culture bioprocesses*. Engineering in Life Sciences, 2015. **15**(5): p. 459–468.
92. Chew, W. and P. Sharratt, *Trends in process analytical technology*. Analytical Methods, 2010. **2**(10): p. 1412–1438.
93. Munson, J., C. Freeman Stanfield, and B. Gujral, *A review of process analytical technology (PAT) in the US pharmaceutical industry*. Current Pharmaceutical Analysis, 2006. **2**(4): p. 405–414.

4 Essential Analytics Framework for PAT Data Analysis

Shawninder Chahal and Xingge Xu

4.1 FUNDAMENTALS OF CHEMOMETRICS

4.1.1 Introduction to Chemometrics

Chemometrics is the science of extracting information from a chemical system using data-driven methods. In bioprocessing, chemometric approaches are primarily applied to predict critical process parameters (CPPs) and critical quality attributes (CQAs) from the data generated by the process analytical technologies (PATs) described in Chapter 3. In addition to prediction, chemometrics plays a key role in anomaly detection, root-cause analysis, and data-driven decision-making to ensure consistent product quality. Two main categories of data typically require chemometric analysis during bioprocessing. The first arises from spectroscopy-based PAT tools, which generate highly multivariate data, often consisting of hundreds or thousands of features corresponding to different wavelengths or frequencies in each spectrum. The second comes from data fusion, where complementary information from multiple PAT tools (e.g., probes and sensors) is integrated to enhance process understanding and monitoring.

In this chapter, we will introduce the essential concepts of chemometrics, including data cleaning, data preprocessing, dimensionality reduction, and basic multivariate modelling.

4.1.2 Measurements

In order to develop a chemometric model, its features need to be defined first. The features will describe the system or phenomenon under observation. A common example of features in a bioreactor can be temperature, pH, dissolved oxygen (DO), and mixing speed, to name a few. Features can also consist of offline measurements, such as glucose concentration measured with an enzymatic assay or viable cell density analyzed with a cell counter. Features can also be categorical; however, they must be modified into a numerical format to be used in a mathematical model. A common approach is *one-hot encoding*, whereby the various categorical possibilities of a feature are encoded into a vector. For example, the impact of three different promoters for gene expression in a model can be expressed as three features: *is_promoter_a*, *is_promoter_b*, and *is_promoter_c*. Running a bioreactor using cells with promoter b would correspond to the following feature values: *is_promoter_a* = 0,

DOI: 10.1201/9781032662961-4

is_promoter_b = 1, and *is_promoter_c* = 0. In the case where a model is used to predict the value of a feature, that feature is often called a target. For example, a capacitance probe is used to measure the viable cell density in a bioreactor. The dataset consists of the following features: capacitance and viable cell density (e.g., measured using a cell counter). In the case where the capacitance is being used to predict the viable cell density, then the viable cell density is often called a target.

4.1.3 Supervised and Unsupervised Learning

Machine learning (ML) is the ability of a machine (e.g., a computer) to learn from data. Although it may not be immediately obvious, there are many parallels to ML and human learning. In the context of PAT, ML algorithms typically fall into two categories: supervised learning and unsupervised learning. The explanation of these categories will follow the analogy of human learning.

Supervised learning consists of ML from data where each sample has one or more target variables that are predicted using the sample's features. Intuitively, humans do this daily. For example, watching movies: after watching a hundred or so movies, a movie enthusiast can *classify* a movie into one or more genres (e.g., action and sci-fi). However, if the hypothetical user has never watched a comedy movie before, the movie fan will find it difficult to classify the first experience of this new category. Likewise, a supervised ML model cannot classify data with labels it has never seen before. Another form of supervised learning involves predicting a continuous variable. After watching several movies and comparing them to the rating provided by movie critics, it may eventually be possible to *predict* their rating (e.g., 7/10 and 8.5/10). So far, the scenarios discussed involve gathering information (e.g., watching movies) and using it to classify (e.g., genre) or predict (e.g., critic rating) some variable. As seen in these two examples, supervised learning algorithms typically fall into two categories: classifier (categorical prediction) and regression (numerical prediction) models. Now, what would happen if this process took place without looking up the genre or critic rating? Would it still be possible to determine underlying patterns from movies watched?

Unsupervised learning consists of ML from data where each sample is unlabeled. For instance, even without previous exposure to the terms "sci-fi movie" or "action movie", it would still be able to implicitly *cluster* movies into different groups. For example, noticing that one set of movies all takes place in outer space, and another set of movies all have high-speed car chases. Similarly, over time (and experience), it would be possible to tell a good movie from a bad one, without knowing the numerical value awarded by the critic rating.

Supervised and unsupervised learning algorithms both play an important role in PAT and are often used together to create end-to-end real-time monitoring pipelines. For example, an unsupervised learning algorithm such as principal components analysis (PCA) can be used to perform dimensionality reduction on a stream of correlated features (e.g., Raman and fluorescence). The reduced dimension dataset can then be used as the input to a supervised learning model to predict the viable cell density in a bioreactor over time. This pipeline would convert bioreactor sensor data into useful information that can be used, e.g., to determine whether the cells are in the growth phase or the stationary phase.

4.1.4 General Steps for Model Development

ML models are rarely as simple as just inputs and outputs. Often, the inputs require extensive data cleaning and preprocessing before even reaching the ML model itself. The choice of data cleaning and preprocessing steps can be crucial to the training of an ML model, especially in the case of model interpretability (i.e., the ability to understand how the model inputs influence the outputs). The general process of model development first consists of data cleaning. At this stage, decisions are made with regard to how missing values will be handled, alignment of time series data coming from unsynchronized sensors, and how (if at all) to interpolate between time points. Once the data is clean, consistent, and complete, it can undergo preprocessing. This typically consists of some combination of scaling (e.g., logarithmic or square root scaling), normalization (setting all samples to the same scale), standardization (setting all features to the same scale), and dimensionality reduction. This chapter discusses each step of this pipeline in further detail.

4.1.5 Matrix, Vector, and Scalar Notation

This chapter and Chapter 5 make extensive use of matrix and vector mathematics. A comprehensive review of linear algebra is beyond the scope of this textbook. However, notation can often vary across the literature, so in this work, the notation for matrices, vectors, and scalar values is defined as follows:

- A matrix is represented by a bold upper-case character, e.g., **M**
- A vector is represented by a bold lower-case character, e.g., **v**
- A scalar is represented by an italicized lower-case character, e.g., *s*
- An element contained in the *i*-th row and *j*-th column of **M** is represented by m_{ij}
- The *k*-th row vector of **M** is represented by $m_{k:}$
- The *k*-th column vector of **M** is represented as $m_{:k}$
- The *k*-th element of vector **v** is represented by v_k

4.2 DATA CLEANING

Data cleaning in the context of PAT typically involves an algorithmic transformation of the raw sensor data in a way that it can be considered consistent (i.e., each sample or time point has the same data structure) and complete (i.e., there are no missing values). Although the exact data cleaning procedure to be used is highly process-specific, this section covers a few scenarios likely to be relevant in PAT.

The application of PAT typically involves data streams from various sources. Moreover, the distance (e.g., in-line, online, and at-line) of an analytical sensor or instrument from the process operation can influence the frequency of the data measurement. Additionally, sensors attached to different software components for data acquisition might not necessarily record data in sync with other software running concurrently. An ML model needs a single matrix, **X**, as an input of size "number of samples" × "number of features". Consider a set of sensors monitoring pH, temperature, and capacitance over the course of a batch run. The pH is recorded every 3 s, the temperature is recorded every 5 s, and the capacitance is recorded every 10 s. In addition, temperature monitoring began 1 s after pH monitoring, and capacitance monitoring began 2 s after pH monitoring. The different sampling periods and start times result in a varying number of samples for each feature at misaligned time points, which becomes problematic when constructing the training matrix (i.e., **X**). Therefore, the incoming data from multiple streams must first be aligned. This will be done in three steps: merging, filling, and slicing. To illustrate a typical data cleaning procedure, consider the first 12 s on a batch run with the aforementioned sensors (Figure 4.1). There are three data streams containing data at different time points.

The first step is to **merge** these three data streams into one matrix. This produces a template for the single **X** matrix that will eventually serve to train the ML model. Solving the problem of three data streams having a different number of samples has introduced a new problem: the resulting matrix is sparse with many missing values.

The second step is to **fill** in the missing values in the new matrix. It may be tempting to perform some sort of interpolation to fill in the missing values (e.g., linear interpolation), but this would lead to training the model on data that would be unavailable in future batch runs. For example, consider the matrix after step 1 in Figure 4.1. From t = 3 s to t = 6 s, the pH increases. However, at t = 4 s, it would be impossible to determine whether the pH was increasing or decreasing, since t = 6 s had not happened yet. In other words, using linear interpolation to fill in the missing values would require having knowledge of the next recorded time point before it happens. In fact, the solution is far simpler. Any missing values can be filled with the last value recorded by the sensor. Once again, this action solves one problem but introduces two new problems. The temperature is undefined at t = 0 s, and the capacitance is undefined at 0 s and 1 s. Moreover, the dataset has grown considerably from 10 datapoints initially to 36 after filling, many of which are repeated values.

Time [s]	pH
0	7.30
3	7.31
6	7.32
9	7.31
12	7.32

Time [s]	Temperature [°C]
1	36.2
6	36.3
11	36.2

Time [s]	Capacitance [pF/cm]
2	10.1
12	10.2

(1)

Time [s]	pH	Temperature [°C]	Capacitance [pF/cm]
0	7.30		
1		36.2	
2			10.1
3	7.31		
4			
5			
6	7.32	36.3	
7			
8			
9	7.31		
10			
11		36.2	
12	7.32		10.2

(2)

Time [s]	pH	Temperature [°C]	Capacitance [pF/cm]
0	7.30	NULL	NULL
1	7.30	36.2	NULL
2	7.30	36.2	10.1
3	7.31	36.2	10.1
4	7.31	36.2	10.1
5	7.31	36.2	10.1
6	7.32	36.3	10.1
7	7.32	36.3	10.1
8	7.32	36.3	10.1
9	7.31	36.3	10.1
10	7.31	36.3	10.1
11	7.31	36.2	10.1
12	7.32	36.2	10.2

(3)

Time [s]	pH	Temperature [°C]	Capacitance [pF/cm]
4	7.31	36.2	10.1
8	7.32	36.3	10.1
12	7.32	36.2	10.2

FIGURE 4.1 Protocol for (1) merging, (2) filling, and (3) slicing data obtained from various sources.

The third step is thus to **slice** the dataset to remove remaining undefined cells in the matrix and to reduce the redundancy introduced in the fill step. In the example shown in Figure 4.1, a time step of 4 s is used. Note that none of the original three data streams has a sampling period of 4 s. In fact, any time step can be used, and this may depend largely on how quickly the bioprocess progresses. In many scenarios, modelling on a 1 s time step may introduce too much noise, and it may be advantageous to slice data to the order of minutes.

The now clean dataset has a unified structure in a single matrix with no missing values and equally spaced time steps. Data expansion was also limited by going from 10 datapoints to only 12 datapoints. Further data preprocessing steps can often be essential to successfully train an ML model.

4.3 DATA PREPROCESSING

The data cleaning process ensures that the data is consistent and complete. Notably, cleaning did not involve any transformation on the actual datapoints themselves, and datapoints were simply rearranged, and time gaps were filled. Recall that an ML model typically needs a single matrix, **X**, as an input with shape "number of samples" × "number of features". Data preprocessing typically involves applying a mathematical function to individual elements or across rows or columns. A data preprocessing pipeline will often contain a sequence of these operations. This section outlines the most common methods typically used in data science applied to PAT.

4.3.1 Function Scaling

Function Scaling is the process of applying a simple mathematical function to each element of a matrix (either the entire matrix or specific columns), typically with the goal of altering its distribution (Equation 4.1).

$$z_{ij} = f\left(x_{ij}\right) \tag{4.1}$$

where f is a function, and **Z** is a matrix containing the scaled data.

Although f can be any function, it is usually a simple function that preserves the order of the values (i.e., sorting the scaled data will result in the same order as sorting the unscaled data). For this reason, logarithmic scaling ($f(x) = \ln(x)$) and power scaling ($f(x) = x^k$) with $k < 1$ are common choices. For power scaling, $k = 0.5$ results in square root scaling. This form of scaling is typically used when values can take a wide range of values covering different orders of magnitude. Moreover, if data follows a log-normal distribution, performing logarithmic scaling will result in data that follows a normal distribution, which is desirable in many ML applications.

4.3.2 Normalization

In large-scale operations, multiple reactors may be running in parallel. Each of these reactors may have various types of spectroscopy measurements being collected (e.g., Raman, fluorescence, and UV-Vis). Although it is important to ensure all instruments and sensors are properly maintained and calibrated, in practice, there may be slight deviations across multiple sensors and over time. Therefore, the overall intensity of spectra may fluctuate across runs and between reactors. Normalization is a method to ensure that data containing multiple related features (e.g., spectra) are on the same scale. There are various methods for performing this normalization, but the most common is L2 normalization as follows:

$$z_{ij} = \frac{x_{ij}}{\sqrt{\sum_{k=1}^{p} (x_{ik})^2}} \tag{4.2}$$

where p represents the number of columns (i.e., features) in $\mathbf{X}$, and $\mathbf{Z}$ is a matrix containing the normalized data. Each row ($z_{i:}$) is now a unit vector.

It is worth noting that normalization inherently results in data loss of one dimension since the magnitude of all samples will be the same. If the spectroscopy was being performed because the key information was contained in the overall intensity, then that information would be lost. However, in most cases, the progress of a bioprocess can be tracked via the ratio of two or more intensities in a spectrum, which will be preserved after normalization. Normalization may also be done before merging the data (e.g., Figure 4.1 – Step 1) since it is done independently for each sample and will not cause any issues with data leakage (to be discussed further in a later section).

4.3.3 Standardization

When handling features with different scales, it may be appropriate to standardize the dataset. For example, consider a matrix $\mathbf{X}$ with features: pH, temperature, and capacitance (Figure 4.1). All these features have different units and scales. Standardization can convert all these features into unitless features with a mean of 0 and a standard deviation of 1 (Equation 4.3):

$$z_{ij} = \frac{x_{ij} - \mu_j}{\sigma_j} \tag{4.3}$$

where μ_j represents the mean of $x_{:j}$, σ_j represents the standard deviation of $x_{:j}$, and $\mathbf{Z}$ is a matrix containing the standardized data.

4.3.4 Robust Scaling

Equation (4.3) works well when the feature is normally distributed. However, if the feature is not normally distributed and/or outliers are present, then robust scaling may be more suitable:

$$z_{ij} = \frac{x_{ij} - m_j}{IQR_j} \tag{4.4}$$

where m_j represents the median of $x_{:j}$, IQR_j represents the interquartile range of $x_{:j}$, and $\mathbf{Z}$ is a matrix containing the robust scaled data.

4.3.5 Rescaling

Another approach to standardization is to have a predefined range over which the process is expected to run. This amounts to rescaling as follows:

$$z_{ij} = \left(d_j - c_j\right)\frac{\left(x_{ij} - a_j\right)}{\left(b_j - a_j\right)} + c_j \tag{4.5}$$

where a_j and b_j represent predefined limits for feature j with the constraint of $a_j < b_j$, c_j and d_j represent the new scale limits for feature j with the constraint of $c_j < d_j$, and $\mathbf{Z}$ is a matrix containing the rescaled data. The parameters a_j and b_j can typically represent the lower and upper limits of a CPP, respectively, or the low and high conditions of a design of experiments dataset, respectively. The parameters c_j and d_j are typically set to −1 and +1, respectively.

4.4 DIMENSIONALITY REDUCTION

Dimensionality reduction plays many roles in ML. In unsupervised learning, it is a great way to easily visualize datasets with many features and quickly determine the existence of clusters or other groupings between similar samples. In supervised learning, it is a useful method to transform and reduce the number of features such that the new set of features is independent from each other. This can lead to faster model training and reduce the risk of overfitting.

4.4.1 Principal Component Analysis

PCA[1,2] is a commonly used tool when working with data containing many features (*i.e.*, dimensions). It is often used as a dimensionality reduction method to visualize high-dimensional data in a 2D or 3D space. However, understanding the mathematical interpretation of a PCA can often reveal hidden patterns in a dataset that may not have been discernible from a simple PCA plot. Most learning resources on the topic of PCA revolve around a demonstration showing data in different groups or categories clustering in different regions of the PCA plot. However, these examples may not always be relevant in the context of time series bioreactor data. This section provides an overview of some data preparation steps and the mathematical interpretation of a PCA, followed by an example relevant to PAT for process monitoring in biomanufacturing.

4.4.1.1 Data Preparation for PCA

The typical approach to representing data for ML operations is in a matrix ($\mathbf{X}$) whereby each row corresponds to a sample, and each column corresponds to a feature. A PCA is centered at the origin. This allows for the interpretation that samples located farther from the origin are also farther from the center of the dataset's distribution. Since there is no bias term in the PCA, the dataset (**X**) must also be centered at the origin. Mean-centering can be done by simply subtracting the mean of each feature (Equation 4.6):

$$z_{ij} = x_{ij} - \mu_j \tag{4.6}$$

where μ_j represents the mean of $x_{:j}$, and **Z** is a matrix containing the mean-centered data.

In the following sections, **Z** can refer to a mean-centered (Equation 4.6) or standardized (Equation 4.3) dataset depending on the context, whereas **X** refers to the original dataset. It is worth noting here that a standardized dataset is also mean-centered, but a mean-centered dataset is not necessarily standardized.

4.4.1.2 Mathematical Interpretation of PCA

The underlying principle of a PCA is to reorient the feature space of a dataset in the directions of greatest variance. This is done by projecting the feature space of a dataset onto a set of orthogonal principal components. The first principal component (PC1) is the unit vector along which the greatest variance in the dataset is captured. The second principal component (PC2) is the unit vector along which the greatest variance is captured in a direction orthogonal to PC1. The third principal component (PC3) is the unit vector along which the greatest variance is captured in a direction orthogonal to PC1 and PC2. From here, the logic extends to PC4, PC5, etc., whereby each additional principal component must capture the greatest amount of variance possible in a direction that is orthogonal to all previous principal components. Another consequence of this is that the principal components are independent features, which can sometimes be a prerequisite for certain statistical or ML models. Mathematically, the PCA transformation for a dataset is represented by a matrix multiplication (Equation 4.7).

$$\mathbf{T} = \mathbf{ZW} \tag{4.7}$$

where **Z** is an $n \times p$ matrix representing the mean-centered dataset, **W** is a $p \times p$ matrix representing the PCA weights, **T** is an $n \times p$ matrix representing the linearly transformed dataset, n is the number of samples, and p is the number of features. Note that when $n \leq p$, the dimensions of these matrices may be different than described here. This is because the number of principal components (p) cannot be greater than $n-1$. For example, consider a dataset with considerably more features than samples. Then a dataset with two samples ($n = 2$) can have 100% of its variance explained by one vector ($p = 1$) pointing from the first sample to the second sample. A dataset with three independent samples ($n = 3$) can always form a plane that contains two orthogonal vectors ($p = 2$). The logic extends to greater sample sizes.

The matrix **T** can be easily understood in the context of a PCA plot. When plotting a PCA using typical software, the first and second columns of **T** would be plotted on the x-axis and y-axis, respectively. In the case of a 3D PCA plot, the third column of **T** would also be plotted on the z-axis. This explains the visualization aspect of PCA; however, it is the **W** matrix which can often reveal further insights, especially in the context of bioprocess monitoring.

The matrix **W** is what defines the PCA for a given dataset. When we think about PCA in the context of ML, it is the **W** matrix that is being *learned* by a *machine* (*e.g.*, a computer). A detailed description of the algorithm that solves for the **W** matrix is beyond the scope of this textbook, but it is typically defined as the eigenvectors of the covariance matrix of **X**, where the *j*-th column of **W** represents the eigenvector of the *j*-th principal component. Considering that each eigenvector has an associated eigenvalue, the fraction of total variance in a dataset that is explained by the *j*-th principal component is defined in Equation (4.8).

$$\text{explained variance ratio} = \frac{\lambda_j}{\sum_{k=1}^{p} \lambda_k} \tag{4.8}$$

where λ_j represents the eigenvalue of the *j*-th principal component (*i.e.*, eigenvector).

4.4.1.3 Applying PCA to the Online Monitoring of Bioreactor Fluorescence

A hypothetical bioprocess will be used to exemplify the role of PCA in process monitoring: a fed-batch cultivation for 72 h, which takes an online fluorescence measurement every 15 min. The excitation wavelength is 365 nm, and the emission spectra are measured from 400 nm to 600 nm in 1-nm intervals (Figure 4.2a). In addition, offline measurements of a compound of interest (COI) are done every 8 h (Figure 4.2b). The bioprocess also has control systems in place to automate the nutrient feed rate such that it remains within a defined concentration interval in the bioreactor.

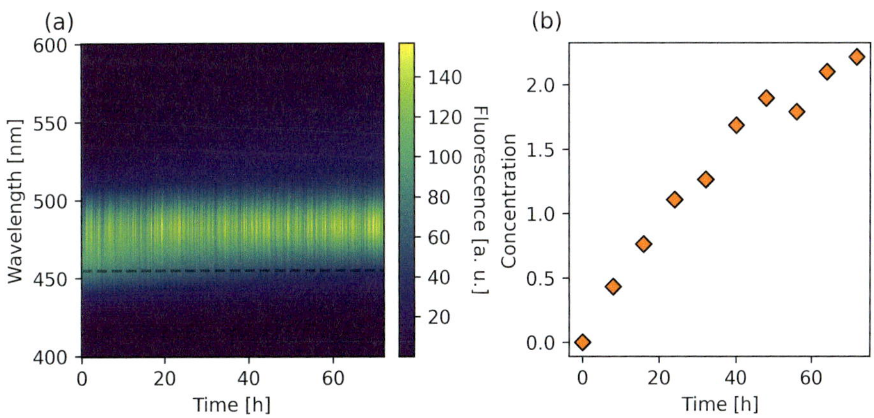

FIGURE 4.2 Hypothetical data of a bioreactor fed-batch process. (a) Online fluorescence measurements taken every 15 min at an excitation wavelength of 365 nm. A dashed black line at 460 nm is shown to emphasize the fluorescence shift over time. (b) Offline measurements of a compound of interest (COI) are taken every 8 h.

Figure 4.3 shows eight snapshots of the fluorescence spectra obtained during a fed-batch run. The objective is to determine if the fluorescence spectra can be used to measure the bioprocess progress. At first glance, there does not appear to be any obvious pattern. The peak is more prominent and taller by the end of the process compared to the start. However, the peak intensity is also seen to oscillate. For instance, starting at 36 h (Figure 4.3e), the peak fluorescence increases at 48 h (Figure 4.3f), then decreases at 60 h (Figure 4.3g), then increases again at 72 h (Figure 4.3h). This is not too surprising since the nutrient feed displays fluorescence in this range, and the control system may cause the fluorescence here to oscillate. There is also a phenomenon happening in the 500 nm–600 nm range where fluorescence is increasing throughout the reactor run. Overall, it is difficult to ascertain what is really happening since there are clearly many fluorescent compounds that are changing in concentration throughout the bioreactor run. Rather than relying on visual inspection of Figure 4.3 alone to determine which wavelengths are of interest, PCA can instead be used.

This example considers a continuous fluorescence spectrum where all the features originate from the same instrument with the same settings. Therefore, Equation (4.6) is used to perform mean centering since the features are provided on the same scale. However, if it happened that working on the same instrument and knowing that two compounds with very different magnitudes of fluorescence intensities were present, but that they should be considered equally important, then it may be appropriate to perform standardization using Equation (4.3). In general, there is no one-size-fits-all approach, and each case should be considered in the context of the data and the intended goals of the PCA.

Unsupervised learning is a major branch of ML that encompasses a variety of algorithms. The key commonality between them is that they can be used to detect

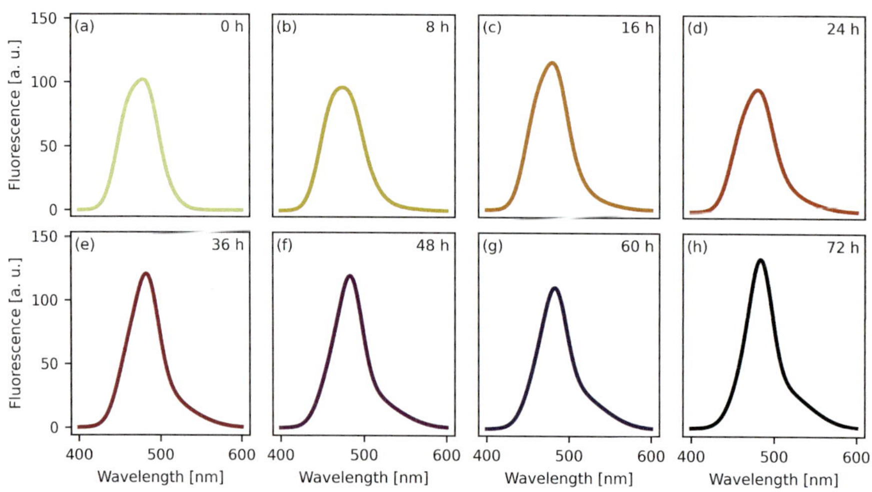

FIGURE 4.3 The fluorescence spectra at (a) 0 h, (b) 8 h, (c) 16 h, (d) 24 h, (e) 36 h, (f) 48 h, (g) 60 h, and (h) 72 h of a hypothetical fed-batch bioprocess.

underlying patterns or trends in a dataset, without knowing the sample labels. This scenario considers an online measurement every 15 min for 72 h, which corresponds to $n = 289$ samples (*i.e.*, timepoints). Fluorescence data is available in intervals of 1 nm from 400 nm to 600 nm, resulting in $p = 201$ features. Therefore, this data can be stored in $\mathbf{X}$, a 289×201 matrix, where x_{ij} represents the fluorescence intensity at time $t_i = \frac{15}{60}(i-1)$ (h) and wavelength $\lambda_j = 400 + (j-1)$ (nm).

4.4.1.4 Scores and Loadings

The PCA scores of the first two principal components (*i.e.*, first two columns of $\mathbf{T}$) are shown in Figure 4.4a. The first and second principal components account for 57.2% and 41.9% of the explained variance (Equation 4.8), respectively. From Figure 4.4a, the diagonal is largely in line with the time-series progression of the bioprocess. Figure 4.4b shows the loadings of the PCA. This represents plotting $\lambda_1 w_{:1}$ and $\lambda_2 w_{:2}$ on the x-axis and y-axis, respectively. Each circle in Figure 4.4a represents a measurement at a timepoint, while each circle in Figure 4.4b represents a feature and its influence on the PCA. Considering the most extreme wavelengths, i.e., 457 nm, 485 nm, and 530 nm, can determine what role they play in the PCA and in the monitoring of the overall bioprocess. The loadings for these three wavelengths are overlaid on Figure 4.4a as well to ease the visualization of their influence.

A high intensity at 530 nm indicates the bioprocess is nearing the end of the run, whereas a high intensity at 457 nm indicates the bioprocess is still at the start of the run. This is also confirmed in Figure 4.4c, which shows the fluorescence at these wavelengths over time. A feature with a loading perpendicular to the progression

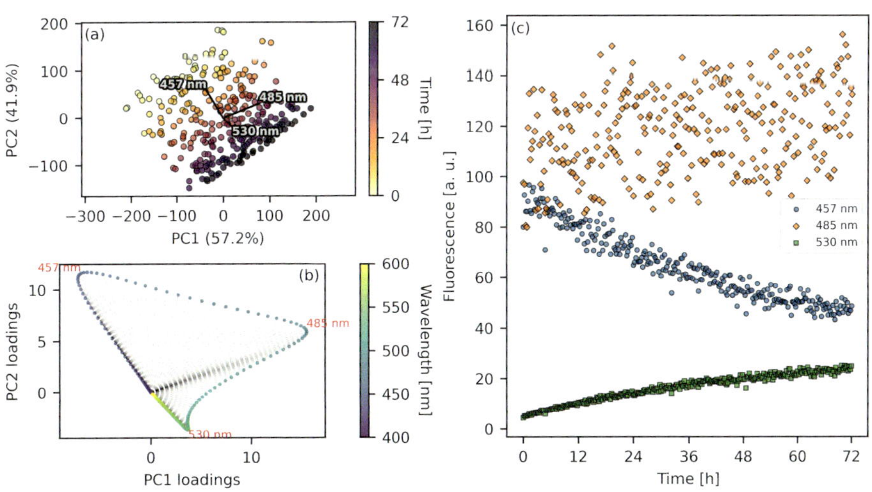

FIGURE 4.4 (a) Principal component analysis of the online fluorescence monitoring data and (a, b) its loadings. Values in parentheses represent the percentage of variance explained by that principal component. Note that in (a), the loadings in (b) were multiplied by a factor of five to facilitate visualization. (c) Change in fluorescence intensity at 457 nm, 485 nm, and 530 nm over the course of the bioreactor run.

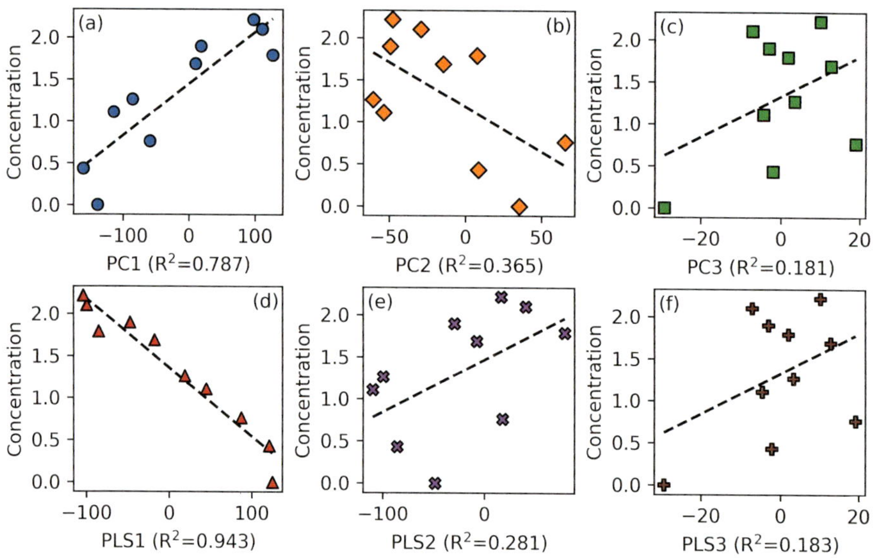

FIGURE 4.6 Comparison of the coefficient of determination (R^2) when predicting the COI concentration using (a) PC1, (b) PC2, (c) PC3, (d) PLS1, (e) PLS2, and (f) PLS3.

REFERENCES

1. Hotelling, H. Relations Between Two Sets of Variates. *Biometrika* **1936**, *28* (3/4), 321–377. DOI: 10.2307/2333955 (acccessed 2025/12/18/). JSTOR.
2. Pearson, K. LIII. On Lines and Planes of Closest Fit to Systems of Points in Space. *The London, Edinburgh, and Dublin Philosophical Magazine and Journal of Science* **1901**, *2* (11), 559–572. DOI: 10.1080/14786440109462720.
3. Wold, S.; Ruhe, A.; Wold, H.; Dunn, I., W. J. The Collinearity Problem in Linear Regression. The Partial Least Squares (PLS) Approach to Generalized Inverses. *SIAM Journal on Scientific and Statistical Computing* **1984**, *5* (3), 735–743. DOI: 10.1137/0905052.

5 Application of Machine Learning in PAT

Shawninder Chahal and Xingge Xu

5.1 OVERVIEW OF MACHINE LEARNING AND ITS RELEVANCE IN BIOPROCESSING

Following the discussion on chemometrics in Chapter 4, machine learning (ML) provides a broader framework for discovering hidden relationships between features and targets. Within the process analytical technology (PAT) context, ML algorithms are applied after appropriate data preprocessing, with or without dimensionality reduction techniques such as principal component analysis (PCA).

ML methods for PAT can be broadly divided into two categories: supervised learning and unsupervised learning. This chapter focuses on supervised learning, which can be further categorized into regression and classification tasks, depending on the intended application. Regression methods are primarily used to predict critical process parameters (CPPs) and critical quality attributes (CQAs). An example introduced in the previous chapter is partial least squares regression (PLSR), which remains a common approach for analyzing spectroscopy-based PAT data. Classification methods are commonly applied for tasks such as anomaly detection.

Most ML models have parameters that are learned from data. These parameters are what the machine learns. Moreover, many ML models have a loss function that is optimized. It is through this optimization process that the parameters can be learned. In general, the actual math behind the optimization itself can be more complex and is beyond the scope of this textbook; therefore, it will not be discussed further.

In this chapter, additional concepts and major ML algorithms relevant to bioprocessing are introduced, and the essential steps in building, validating, and maintaining ML models for PAT applications are discussed.

5.2 BIAS–VARIANCE TRADE-OFF

When training a ML model, it is important to ensure that the model is learning the underlying phenomenon in your data and not just fitting to the noise. For instance, if you have data that shows curvature, you may decide to fit your data to a quadratic formula ($y = ax^2 + bx + c$). You will note that you can also fit your data to a cubic formula ($y = ax^3 + bx^2 + cx + d$) to get a better R^2. In fact, a higher order polynomial will always generate a higher R^2 on the training data. However, high-order polynomials will tend to oscillate. Is your bioprocess really fluctuating so frequently, or did your model fit to the noise in your data instead of your actual phenomenon? When a model learns the variance in the data instead of the underlying phenomenon, this is referred to as overfitting. Conversely, if your data was curved and you fit it to a

DOI: 10.1201/9781032662961-5

linear equation ($y = ax + b$), you may notice a large bias error in your model. This is referred to as underfitting.

In addition to having parameters, most ML models also have hyperparameters that influence *how* the machine learns the parameters. Many of these hyperparameters serve to control the regularization applied when a model learns parameters, i.e., to ensure the model does not underfit or overfit. Regularization can come in many forms, and these hyperparameters will be discussed in the individual model subsections. Whereas parameters are learned by the model by fitting to the data, hyperparameters cannot typically be learned. The process of hyperparameter optimization is discussed in Section 5.6.

5.3 REGRESSION

Regression models have numerical inputs and outputs and play a key role in PAT, especially in the field of real-time process monitoring.

5.3.1 ORDINARY LEAST SQUARES REGRESSION

Ordinary least squares (OLS) regression is one of the simplest regression models. It has no hyperparameters, and it follows a simple linear regression (Equation 5.1).

$$\hat{\mathbf{Y}} = \mathbf{XW} \tag{5.1}$$

where $\hat{\mathbf{Y}}$ is a matrix of shape ($n \times q$) representing the *predicted* targets, $\mathbf{X}$ is a matrix of shape $\left(n \times (p+1)\right)$ representing n samples with p input features and an additional column of 1 s (i.e., $x_{i,(p+1)} = 1$) for a bias term, $\mathbf{W}$ is a matrix of shape $\left((p+1) \times q\right)$ representing the weights of the OLS model. The $\mathbf{W}$ matrix represents the parameters of the model. When the features in $\mathbf{X}$ are all on the same scale (*e.g.*, *via* standardization), then w_{ij} can also represent the influence of the i-th feature on the prediction of the j-th target. The $\mathbf{W}$ matrix is learned from the training data by optimizing an OLS loss function (Equation 5.2).

$$L(\mathbf{W}) = \sum_{i=1}^{n} \sum_{j=1}^{q} \left(y_{ij} - \sum_{k=1}^{p+1} x_{ik} w_{kj} \right)^2 \tag{5.2}$$

where y_{ij} represents the *actual experimental value* of the i-th sample and j-th target, and L is the loss to be minimized by finding the optimal values of $\mathbf{W}$.

Although OLS is a linear model, additional features can be engineered that are nonlinear in nature but may appear linear to the loss function. For example, consider a design of experiment (DOE) dataset that has four independent features: pH, temperature, stirring speed, and time. Then a fifth column could be generated that is equal to $(\textit{temperature})^2$ as a quadratic term, and a sixth column that represents $(\textit{temperature})(\textit{time})$ as an interaction term. However, doing so introduces multicollinearity and would mean that the features are no longer independent. Fortunately, there are alternative regression methods that can take multicollinearity into account.

5.3.2 Ridge Regression

Ridge regression[1] is another form of linear regression that is similar to OLS but adds an additional term to the loss function to minimize the impact of multicollinearity (Equation 5.3).

$$L(\mathbf{W}) = \sum_{i=1}^{n}\sum_{j=1}^{q}\left(y_{ij} - \sum_{k=1}^{p+1} x_{ik}w_{kj}\right)^2 + \alpha\sum_{k=1}^{p+1}\sum_{j=1}^{q}\left(w_{kj}\right)^2 \tag{5.3}$$

where α is a hyperparameter with a positive value representing the regularization term. Since α is a hyperparameter, it must be set by the user and cannot be learned by the model. Finding the optimal value of α is the topic of Section 5.6. It can be readily seen that when $\alpha = 0$, the ridge regression will find the same weights (**W**) as the OLS model. The higher the value of α, the more the ridge regression optimizer will try to minimize the squared values in **W**. It also follows that when $\alpha \to \infty$, then $w_{ij} \to 0$, and will lead to underfitting. In other words, α can be increased to reduce overfitting. It should also be stated that it is ideal to use features that are on the same scale and centered at zero so that the values of **W** are also on the same scale.

Let's explore how minimizing **W** can help reduce the impact of multicollinearity. Consider a fluorescence spectrum where we have a compound that exhibits peak fluorescence at a wavelength of 501 nm. Our instrument was only measuring the fluorescence every 2 nm, and so our closest points of evaluation are 500 nm and 502 nm, in addition to every 2 nm increment before and after, in the 400 nm–700 nm range. If the fluorescence spectrum has no skew to it, then the fluorescence at 500 nm and 502 nm should be the same in all our samples. In other words, these two features are strongly correlated. In practice, there will always be some degree of noise, meaning the two features will not be exactly the same. However, after collecting many samples, their mean fluorescence should be almost identical. There are a few problems that could arise with using an OLS model. As the model iteratively minimizes the loss function (Equation 5.2), it may first put more importance on the 500-nm feature. Then in the next iteration, it may place more importance on the 502-nm feature. It may then keep switching back and forth until settling on whichever one was more important in the iteration when the loss function was sufficiently minimized. Ultimately, the model may settle on completely different weights for the feature at 500 nm and 502 nm. In some cases, this may not actually impact model performance. However, if the role of each feature is to be interpreted in the model, the results may be confusing. A scenario may arise where it is determined that a high intensity at 500 nm implies a high concentration, but a low intensity at 502 nm implies a high concentration as well. An OLS model may, for example, produce coefficients as shown in Equation (5.4).

$$c = 10x_{500} - 4x_{502} \tag{5.4}$$

where c represents the concentration, and x_{500} and x_{502} represent the fluorescence intensity at 500 nm and 502 nm, respectively. In this example, 10 and –4 are hypothetical weights for x_{500} and x_{502}, respectively. If we know that x_{500} is approximately equal to x_{502}, then we know that as long as their coefficients have the same sum (*i.e.*, $10-4=6$), then the prediction for c will remain the same. If we use $\alpha=1$ in the ridge regression loss function (Equation 5.3), the regularization term using the coefficients from the OLS model (Equation 5.4) equals 116 (Equation 5.5):

$$\alpha\left(\left(w_{500}\right)^2+\left(w_{502}\right)^2\right)=(1)\left(\left(10^2\right)+\left(-4\right)^2\right)=116 \tag{5.5}$$

However, the ridge regression will try to minimize this. A potential solution the ridge regression optimization could reach would be to make the coefficients equal while still having a sum of 6. Equation (5.6) shows how this model would look, while Equation (5.7) shows how the loss function has been greatly optimized (*i.e.*, reduced from 116 to 18) by having equal coefficients for the correlated terms.

$$c=3x_{500}+3x_{502} \tag{5.6}$$

$$\alpha\left(\left(w_{500}\right)^2+\left(w_{502}\right)^2\right)=(1)\left(3^2+3^2\right)=18 \tag{5.7}$$

The coefficients of the ridge regression model now better reflect the underlying phenomenon that the fluorescence intensity at 500 nm and 502 nm have an equal impact in predicting the concentration, and that a higher intensity of both implies a higher concentration. Moreover, consider the fact that this logic likely extends further down the spectra as well. For instance, 498 nm and 504 nm may also have similar fluorescence, and so on.

It is important to bring attention to the role of the hyperparameter α in the ridge regression loss function. This hyperparameter needs to be carefully selected, since too low a value will essentially yield little difference from an OLS model, but too high a value will cause all the weights to approach zero at the expense of model performance.

5.3.3 Principal Component Regression

Principal component regression (PCR) is an extension of unsupervised PCA. In essence, it is an OLS model that is trained on the principal components instead of the original features. The hyperparameter is the number of principal components to use. Typically, this is considered in terms of the fraction of variance captured by the principal components. For example, if you wanted to keep the principal components that account for 95% of the total variance in the dataset, then you can keep removing the last principal component until the sum of the explained variance of the remaining principal components is less than 95%. It is important to note here that variance is not synonymous with error or noise. After all, the majority of the variance in your dataset should come from the underlying phenomena in your data. The theory behind this approach is that the first few principal components will be more likely to contain the variance associated with the underlying phenomena in your data, and the last few principal components will just contain the variance associated with noise. The other benefit of this approach is that the PCA will handle the issue of multicollinearity in your data as well. Therefore, when the weights from the PCA are matrix multiplied by the weights of the OLS model, they will produce weights for the original features that will produce more reliable feature importance measurements

5.3.4 Partial Least Squares Regression

PLSR[2] is covered more in-depth in Section 4.4.2. The hyperparameter is the number of components to use, which is similar to PCR, and controls the amount of variance included in the model. The choice between performing PLSR vs PCR can depend on the goals of the model. If the goal is to create a model using the fewest orthogonal components possible, then PLSR will typically do a better job than PCR since PLSR's dimensionality reduction is supervised. On the other hand, if the goal is just to make a well-optimized linear model on highly correlated features (and the number of components used is not important), then one may find that both PCR and PLSR will often have similar performance. Another consideration may be that some software packages have PLSR built in due to its popularity, whereas PCR may have to be implemented manually.

5.3.5 *k*-Nearest Neighbors Regression

The regression models covered up to this point have all been linear in nature. Even if we derive nonlinear features such as x^2, the underlying model still takes the form of Equation (5.1) and requires the user to know what the relationship between the targets and features is (e.g., is the observed curve quadratic, cubic, or exponential?). Many nonlinear models exist; however, k-nearest neighbors[3] regression is a robust model that does not make any underlying assumptions about the distribution of the input features and their relationship to the target variable. Its operation is quite simple, consider a training dataset consisting of a matrix $\mathbf{X}$ of shape n samples $\times$ p features (e.g., pH and size) and a vector $\mathbf{y}$ of size n samples containing a target value to be predicted (e.g., drug efficacy). Now consider that you have a set of new samples in a matrix $\mathbf{U}$ of shape m samples $\times$ p features for which you want to predict and store the target values in a vector $\mathbf{v}$ of size m samples.

The first step is to calculate a distance matrix $\mathbf{D}$ of shape $m \times n$ using Equation (5.8):

$$d_{ij} = \sqrt{\sum_{k=1}^{p} \left(u_{ik} - x_{jk}\right)^2} \tag{5.8}$$

where d_{ij} represents the Euclidean distance between the j-th sample in the training matrix $\mathbf{X}$ and the i-th sample in matrix $\mathbf{U}$ for which we want to make a prediction. Therefore, it would typically require that the features be on the same scale using a method such as standardization. Alternatively, there is no requirement that the Euclidean distance be used. For example, UniFrac is commonly used to measure the distance between microbial communities,[4] and can therefore also be used to construct a distance matrix.

The next step is to perform our prediction of **v** (Equation 5.9):

$$v_i = \sum_{j=1}^{n} a_{ij} w_{ij} y_j \tag{5.9}$$

where **A** is an activation matrix and **W** is a weights matrix. These two matrices will be constructed based on our two hyperparameters. The first is the k in k-nearest neighbors. This value is a positive integer and represents the number of samples in the training dataset that are closest to a given sample in the predicting dataset that will be used in the prediction. It influences the structure of **A** according to Equation (5.10):

$$a_{ij} = \begin{cases} 1, & d_{ij} \le \left(k^{th} \text{ smallest value in } d_{i:}\right) \\ 0, & d_{ij} > \left(k^{th} \text{ smallest value in } d_{i:}\right) \end{cases} \tag{5.10}$$

The second hyperparameter is the type of weighting structure to be used, which will influence the structure of **W**. The most common are uniform weights (Equation 5.11) and distance-based weights (Equation 5.12):

$$w_{ij} = \frac{1}{k} \tag{5.11}$$

$$w_{ij} = \frac{1}{d_{ij}} \tag{5.12}$$

Let us consider how these two hyperparameters influence our model in terms of the bias–variance trade-off. If k is reduced, then the model will move more toward overfitting, and if k is increased the model will move toward underfitting. This becomes intuitive by thinking of the extreme scenarios. When $k = 1$ we are essentially predicting an unknown sample to be equal to whatever the closest sample is in the training set. If there are any outliers, this will cause a lot of noise for predictions in the vicinity of that training sample. Conversely, if $k = n$ then we are essentially taking the average over the entire dataset, which will cause a constant prediction when using uniform weights. Next, let us consider the weighting scheme. Uniform weights should be more robust to outliers, especially as k increase. For example, consider, $k = 10$, even if your new sample is close to an outlier in the training set, that training sample will only have a weight of 10%, and ideally, the other nine samples can bring the prediction back to a more reasonable value. Conversely, when using distance weights, the weight of that outlier might still be very large. However, in regions of the feature space where there are not many training samples, distance-based weighting can be useful; when there are, for example, eight samples nearby and two samples very far away, we may not want to include those two samples in our prediction, so they will have a very low weight due to their distance. Finally, the choice of distance matrix D can also be considered a third hyperparameter if there are multiple options available for a given dataset. However, this is often chosen based on what is appropriate given the type of data being used, rather than something that is optimized for regularization purposes.

5.4 CLASSIFICATION

Classification is another major branch of supervised learning. Whereas regression allows one to predict a continuous numerical variable, classifiers are typically used to predict a categorical variable, especially those where there is no numerical association or order to the categories. For example, if you took a series of medical diagnostic tests to screen for a disease, you could make a classifier model to make that prediction based on the data collected from these diagnostic tests. Some classifier models also provide a probability measure on the prediction.

Classification problems fall under two categories: binary classification and multiclass classification. Binary classification seeks to discern which of two classes (e.g., disease, no disease) the data's features describe. In binary classification, there is often a *positive* class and a *negative* class. The choice of which class is positive and which is negative is arbitrary. For example, for a disease detection model, the presence of a disease may be considered a positive condition (e.g., someone tested positive for a disease); however, it can just as easily be labeled as a negative (i.e., it is a negative outcome to have the disease). It will only require you to invert the interpretation of your model (e.g., there is a 35% probability that a patient has a disease vs. there is a 65% probability that a patient does not have a disease). Multiclass classifiers seek to discern which of more than two classes (e.g., blood type, O–, O+, and A–) the data's features describe. Depending on the model being used, the algorithm may be different when dealing with multiclass problems, and there may be special considerations to take when developing your classifier model.

The major applications of classification methods in PAT are fault detection and diagnosis, and batch comparison. Classifiers such as random forests, artificial neural networks, and support vector machines (SVMs) have been employed to detect anomalies in aeration, agitator power, and feed rate during bioreactor operation by analyzing combined sensor signals and control outputs.[5] In addition, hierarchical clustering methods have been reported for distinguishing between high-performing and low-performing batches at an early stage of cell growth in bioreactors.[6]

5.4.1 Logistic Regression

Although the name would suggest that this model is both nonlinear and a regression model, the fact is that logistic regression is often used as a linear ML classifier model. To understand how, let us look at the equation that defines logistic regression.

Logistic regression for binary classification works by applying the standard logistic function (Equation 5.13) to the output of a linear model (Equation 5.14):

$$\hat{y}_i = \frac{1}{1 + e^{-z_i}} \tag{5.13}$$

$$\mathbf{z} = \mathbf{X}\mathbf{w} \tag{5.14}$$

where $\hat{\mathbf{y}}$ is a vector of size n representing the *predicted* probability that the i-th sample belongs to the positive class, $\mathbf{X}$ is a matrix of shape $(n \times (p+1))$ representing n samples with p input features and an additional column of 1s (i.e., $x_{i,(p+1)} = 1$) for a bias term, $\mathbf{w}$ is a vector of size $(p+1)$ representing the weights of the linear component of the model, and $\mathbf{z}$ is a vector of size n that represents the output of the linear component of the model and the input to the logistic component. Logistic regression is trained by minimizing the loss function shown in Equation (5.15):

$$L(\mathbf{w}) = \sum_{i=1}^{n} \left(-y_i \ln(\hat{y}_i) - (1 - y_i)\left(\ln(1 - \hat{y}_i)\right)\right) + \frac{1}{C}\sum_{j=1}^{p+1} (w_j)^2 \tag{5.15}$$

where $\mathbf{y}$ is a vector of size n representing the *actual* probability that the i-th sample belongs to the positive class. In most cases, this will mean $y_i = 1$ if the i-th sample belongs to the positive class and $y_i = 0$ if the i-th sample belongs to the negative class. The loss function also makes clear the role of the hyperparameter C, which is an inverse regularization term. Higher values of C will cause the model to overfit, whereas lower values will cause it to underfit.

Multiclass classification via logistic regression is performed by training a series of one-vs-rest models. For example, consider a test for classifying a sample as belonging to one of three drug formulations, i.e., drug A, B, and C. Model 1 will be trained such that drug A belongs to the positive class and drugs B and C belong to the negative class. Model 2 will be trained such that drug B belongs to the positive class and drugs A and C belong to the negative class. Model 3 will be trained such that drug C belongs to the positive class and drugs A and B belong to the negative class. Each of these models will have its own $\mathbf{w}$ vector that can be concatenated into a matrix $\mathbf{W}$, such that $w_j = \mathbf{w}$ for the j-th model. Similarly, this will result in a matrix $\mathbf{Z}$ such that $z_j = \mathbf{z}$ for the j-th model. In the example with drugs A, B, and C, we had three classes. Let us consider the generic scenario with q classes. The logistic regression model for multiclass classification becomes Equation (5.16):

$$\hat{y}_{ij} = \frac{e^{z_{ij}}}{\sum_{k=1}^{q} e^{z_{ik}}} \tag{5.16}$$

where $\hat{y}$ is a matrix of shape $(n \times q)$ representing the *predicted* probability that the i -th sample belongs to the j-th class.

5.4.2 Partial Least Squares–Discriminant Analysis (PLS-DA)

Partial least squares–discriminant analysis (PLS-DA) uses PLSR as a linear model and passes it through an activation function to add discriminant analysis to the PLSR to make a classification decision. In binary classification, a PLSR model is trained where $y_i = 1$ if the i-th sample belongs to the positive class and $y_i = -1$ if the i-th sample belongs to the negative class. When making a prediction, the discriminant analysis is simple: if $\widehat{y_i} > 0$ then the i-th sample is predicted to belong to the positive class, then it is predicted to belong to the negative class. The multiclass case is similar to logistic regression. Multiple one-vs-rest models are trained, where each class has an opportunity to be the positive class. Then the model with the highest value of $\widehat{y_i}$ determines the prediction for the i-th sample.

5.4.3 Support Vector Machine

SVMs are unique and flexible ML classifier models. SVMs specialize in training on data that would otherwise not be separable by linear models by performing what has come to be known as the kernel trick. Although the math behind SVMs and the kernel trick is beyond the scope of this textbook, it can be viewed as a computationally efficient method to linearize a nonlinear separation problem. This is done by using a kernel function to measure the similarity between new samples and previously seen samples. A variety of kernel functions exist; however, the most commonly used is the radial basis function (RBF) kernel, which assigns a similarity score of 0 when two samples are infinitely distant from each other, and a similarity score of 1 when two samples are exactly the same. Another unique aspect of SVMs is their use of the hinge loss function, which places greater emphasis on samples at the frontier of different clusters instead of considering the entire distribution as a whole.

5.4.4 *k*-Nearest Neighbors Classifier

The classification version of k-nearest neighbors[3] works similarly to the regression version. The difference is in the interpretation of vectors **v** and **y** in Equation (5.9). The vector **y** of size n samples now contains the target *category* to be predicted instead of a numerical value. Similarly, the coefficients $a_{ij}w_{ij}$ represent the number of votes this category received in the prediction. For example, consider a hypothetical scenario in the case of three types of drug formulations, A, B, and C. The hypothetical summation in Equation (5.9) with $k = 3$ and distance-based weights may look something like this (Equations 5.17 and 5.18):

$$v_i = \sum_{j=1}^{n} a_{ij} w_{ij} y_j = (1)(2.12)(B) + (1)(1.24)(C) + (1)(0.57)(C) + (0)(0.24)(A) + \dots \tag{5.17}$$

$$v_i = \{0\ A,\ 2.12\ B,\ 1.81\ C\} \tag{5.18}$$

The conclusion of this hypothetical prediction is that the i-th sample is predicted to be drug B since it had 2.12 votes, which was the highest of any of the classes.

5.4.5 Gaussian Naïve Bayes

Gaussian naïve Bayes algorithms[7] assume that for each class, each feature follows a normal distribution. Let **M** and **S** be $p \times q$ matrices where p is the number of features, q is the number of classes, and m_{ij} and s_{ij} represent the mean and standard deviation, respectively, of the i-th feature for training samples belonging to the j -th class. Now consider that you have a set of new samples in a matrix **U** of shape

m samples × p features for which you want to store their predicted class in a vector **v** of size m samples. The prediction begins by calculating the likelihood of the sample belonging to each class and storing these likelihoods in a matrix **R** of shape m samples × q classes (Equation 5.19):

$$r_{kj} = b_j \prod_{i=1}^{p} \frac{1}{\sqrt{2\pi\left(s_{ij}\right)^2}} e^{-\left(\frac{\left(u_{ki}-m_{ij}\right)^2}{2\left(s_{ij}\right)^2}\right)} \tag{5.19}$$

where b_j represents the prior probability of the j-th class in the training dataset. This is simply the probability of a sample belonging to a class without any knowledge of its features. In other words, this means that b_j is simply the fraction of training samples belonging to the j-th class. It then follows that the prediction is simply the class with the highest likelihood (Equation 5.20):

$$v_k = \text{argmax}\left(r_{k:}\right) \tag{5.20}$$

where **v** contains the predicted index of the k-th sample. For example, if $v_k = 2$, then the k-th sample is predicted to belong to the second class.

Gaussian naïve Bayes is particularly interesting in that it has no hyperparameters. However, it must be reiterated that it is intended to work with data that is normally distributed, which is reflected in the structure of Equation (5.19).

5.4.6 Decision Tree Classifier

The structure of a decision tree classifier is quite different from the models we have discussed up until this point. It is not easily described as a series of functions applied to vectors and matrices. Instead, as the name suggests, it follows the structure of a decision tree or flow chart. Consider a sample for which a prediction is to be made. At each node of the decision tree, the value of a single feature is considered. If this value is above a specified threshold, then it goes to the right path, otherwise it goes to the left path. At the next node, the same process occurs with a single feature and a specified threshold. This process continues until a final prediction is made. The value of the threshold at each node is a parameter to be learned by the model from the data by choosing a threshold that will minimize the Gini impurity in the child nodes. There are several methods to control overfitting of decision trees, a common hyperparameter being to control the maximum depth of the tree. Reducing the maximum depth will reduce the degree of overfitting.

5.4.7 Random Forest

Much like a forest contains many trees, a random forest contains many decision trees. The "random" component comes from the fact that each decision tree is trained on a random subset of the features and samples found in the overall training dataset. The prediction from each decision tree becomes its vote for the overall random forest prediction. For example, if a random forest contains 100 decision trees with 70 trees predicting class A and 30 trees predicting class B for a sample, then the random forest will predict class A. Random forests are very robust against overfitting. Generally, the main hyperparameter is the number of decision trees to use. Increasing the number of trees generally reduces overfitting asymptotically, after which there will be negligible difference in performance, but training time will increase significantly. Other hyperparameters include the number of samples and features used to train each decision tree.

5.5 ARTIFICIAL NEURAL NETWORKS

Artificial neural networks are a diverse class of ML models that can be used for both classification and regression (and other methods as well). Neural networks typically consist of a series of linear models that are then passed through a nonlinear activation function to create hidden layers. The output of the first hidden layer becomes the input of the second hidden layer, the output of the second hidden layer becomes the input of the third hidden layer, and so on until the output layer is reached. Each node in a hidden layer consists of a linear model followed by a nonlinear activation function. This allows neural networks to learn complex patterns efficiently. However, the complexity of neural networks requires a deep understanding of how they operate to properly optimize them, since they can be prone to overfitting, especially on smaller datasets. The mathematics of neural networks along with their various hyperparameters are beyond the scope of this textbook. However, we will introduce some major neural network architectures so that the reader may know where to look if a certain model may fit their application.

5.5.1 Fully Connected Neural Network

The most basic type of neural network is the fully connected neural network (FCNN). They can be used for both regression and classification tasks and can take on a variety of loss functions to achieve different goals. They are well-suited for modeling when the underlying patterns in the data are completely unknown. However, due to their fully connected nature, these models can grow in their number of parameters very quickly, making them prone to overfitting if the sample size is small. Moreover, this model treats the features and samples as being independent. This may not be the case, e.g., with time series spectroscopy data, where the features may be highly correlated. For instance, the spectra of a sample at the current time point are related to the spectra of the same sample at the previous time point. Next, let's explore two models that are well-suited for these cases.

5.5.2 Convolutional Neural Network

Convolutional neural networks (CNNs) seek to overcome two of the shortcomings of FCNNs. Whereas FCNNs treat all features independently, CNNs require the features to be ordered. For instance, a Raman or nuclear magnetic resonance (NMR) spectra may have sharp peaks, but the peak itself will still be described over a range of wavenumbers (e.g., two sharp peaks were observed in the 800–900 cm^{-1} range). Moreover, there are many ways these peaks can appear; perhaps the left peak is taller than the right peak. An FCNN will have no concept of a peak, because it does not know that the feature at 800 cm^{-1} is located next to the feature at 801 cm^{-1}. Therefore, the FCNN cannot recognize the shape of the peak, which could contain valuable information for identifying chemical structures. A CNN is designed to both recognize patterns where the features are ordered while simultaneously reducing the number of overall parameters needed by sharing weights between features. The CNN scans the spectra, applying a linear transformation of the spectra followed by a nonlinear activation function. Conceptually, during training, the CNN will learn which types of peaks are worth amplifying in each layer and which should be suppressed (since they are irrelevant to the prediction task). The CNN can also be applied in a 2D context, for example, with a complete fluorescence spectrum across a range of excitation and emission wavelengths.

5.5.3 Recurrent Neural Network

We have seen how CNNs can be used to develop models where the features are in an ordered structure. However, although the features are ordered in, for example, a fluorescence spectrum, there is no concept of forward or reverse. Whether the spectrum is scanned from 400 nm to 700 nm or 700 nm to 400 nm, the same transformation would be applied when using a CNN. Therefore, the features are ordered, but they were not sequential. The most common type of sequential data in PAT is time series data. For instance, if we have developed a model for real-time glucose monitoring, and we want to predict what the glucose concentration will be sometime in the future, we can train an RNN to discover the sequential trend in the glucose concentration at the next few time points. In particular, RNNs are well-suited to discover nonlinear time series trends, based on a variety of input data (i.e., sensors). Many types of RNN architectures exist, but one of the most well-known is long short-term memory (LSTM).

5.6 HYPERPARAMETER OPTIMIZATION

In the previous section, we saw that most ML models have hyperparameters that can be tuned to control overfitting. A common method for optimizing these hyperparameters is grid search cross-validation. The grid search component is simple; consider a model such as a k-nearest neighbors classifier, which has k (i.e., the number of neighbors) and weight type (i.e., uniform or distance-based) as hyperparameters. To find the best combination of these two hyperparameters, you may set up a grid with every combination within a selected range. An example is shown in Table 5.1. Note that a grid search does not require there to be two hyperparameters. There could be as many as are relevant to the task; however, training time will increase with each dimension. A search can also be done using one hyperparameter.

The next step is to perform cross-validation on each of these combinations to determine how they impact the performance of a model when tested on data that was not involved in the training process. Cross-validation is performed by splitting

TABLE 5.1
An Example of a Grid Search Using a *k*-Nearest Neighbors Classifier

k	1	3	5	7	9	11	13
Uniform weights	$k = 1$ Uniform weights	$k = 3$ Uniform weights	$k = 5$ Uniform weights	$k = 7$ Uniform weights	$k = 9$ Uniform weights	$k = 11$ Uniform weights	$k = 13$ Uniform weights
Distance weights	$k = 1$ Distance weights	$k = 3$ Distance weights	$k = 5$ Distance weights	$k = 7$ Distance weights	$k = 9$ Distance weights	$k = 11$ Distance weights	$k = 13$ Distance weights

a dataset into different groups called folds and alternating between which folds are used for training, and which are used for validation. A common number of folds is five or ten. To understand why this separation of training and validation samples is important, consider the following example. Imagine you are taking a course, and your professor gives you practice problems to help you study for an upcoming exam. Now, when you get to the exam, you obviously will not see the same practice problems. That is because memorizing the solutions to a set of problems does not tell the professor whether you learned the underlying concepts. Therefore, the professor decides to give you a new set of problems on the exam that cover the same topics and concepts of the practice problems instead. Similarly, ML models need to be validated on data that was not involved in their training, otherwise they could easily attain 100% accuracy depending on the model. For example, *k*-nearest neighbors with k=1 would have a 100% accuracy if the training samples were used to validate the model, since the 1-nearest neighbor will be the very same sample that you are testing (since that sample also appears in the training dataset).

Returning to our grid search example, consider a dataset containing blood test data being used to predict the presence of a disease. If the disease has a prevalence of 10% and there are 1000 samples, then 900 samples would be healthy patients and 100 samples would belong to sick patients. In a stratified five-fold cross-validation, each fold would contain 180 samples from healthy patients and 20 samples from sick patients. These five folds could simply be labelled fold 1, 2, 3, 4, and 5. Considering our grid search in Table 5.1, the first step would be to train a *k*-nearest neighbors classifier model using k=1 and uniform weights on the samples in folds 2, 3, 4, and 5. Then this model can be tested on fold 1 to get an accuracy measurement. Next, the model will be trained on folds 1, 3, 4, and 5, then tested on fold 2 to get an accuracy measurement. This is repeated until each fold has an opportunity to be the test set. This completes the cross-validation for the hyperparameter combination k=1 and uniform weights.

Cross-validation as described in the previous paragraph will then be performed on a *k*-nearest neighbors classifier model using k=3 and uniform weights. Then it will be performed on all the hyperparameter combinations in Table 5.1, and the hyperparameter pair with the highest mean accuracy will be chosen. Table 5.2 shows an example of the results of a hypothetical grid search cross-validation. Based on the results of this grid search cross-validation, we conclude that k=5 and distance-based

TABLE 5.2
An Example of the Results of a Grid Search Cross-Validation Using a *k*-Nearest Neighbors Classifier

k	1	3	5	7	9	11	13
Uniform weights	Fold 1: 83.5%	Fold 1: 81.0%	Fold 1: 86.0%	Fold 1: 86.5%	Fold 1: 82.0%	Fold 1: 78.5%	Fold 1: 68.0%
	Fold 2: 90.0%	Fold 2: 78.5%	Fold 2: 82.5%	Fold 2: 83.0%	Fold 2: 82.0%	Fold 2: 80.5%	Fold 2: 76.0%
	Fold 3: 94.5%	Fold 3: 77.0%	Fold 3: 78.0%	Fold 3: 82.5%	Fold 3: 79.5%	Fold 3: 84.0%	Fold 3: 82.5%
	Fold 4: 91.0%	Fold 4: 79.0%	Fold 4: 75.5%	Fold 4: 86.5%	Fold 4: 82.5%	Fold 4: 78.0%	Fold 4: 85.5%
	Fold 5: 84.5%	Fold 5: 84.0%	Fold 5: 76.5%	Fold 5: 94.0%	Fold 5: 87.0%	Fold 5: 74.5%	Fold 5: 74.5%
	Mean: 88.7%	Mean: 79.9%	Mean: 79.7%	Mean: 86.5%	Mean: 82.6%	Mean: 79.1%	Mean: 77.3%
	SD: 4.1%	SD: 2.4%	SD: 4.0%	SD: 4.1%	SD: 2.4%	SD: 3.1%	SD: 6.2%
Distance weights	Fold 1: 84.5%	Fold 1: 84.0%	Fold 1: 87.5%	Fold 1: 82.0%	Fold 1: 77.5%	Fold 1: 81.0%	Fold 1: 70.0%
	Fold 2: 84.0%	Fold 2: 84.0%	Fold 2: 94.5%	Fold 2: 81.0%	Fold 2: 85.0%	Fold 2: 75.0%	Fold 2: 64.0%
	Fold 3: 87.0%	Fold 3: 72.5%	Fold 3: 91.0%	Fold 3: 80.0%	Fold 3: 91.5%	Fold 3: 75.0%	Fold 3: 72.5%
	Fold 4: 87.5%	Fold 4: 71.5%	Fold 4: 93.0%	Fold 4: 82.5%	Fold 4: 95.5%	Fold 4: 78.5%	Fold 4: 70.0%
	Fold 5: 85.0%	Fold 5: 65.5%	Fold 5: 97.5%	Fold 5: 83.5%	Fold 5: 97.0%	Fold 5: 72.0%	Fold 5: 71.5%
	Mean: 85.6%	Mean: 75.5%	Mean: 92.7%	Mean: 81.8%	Mean: 89.3%	Mean: 76.3%	Mean: 69.6%
	SD: 1.4%	SD: 7.3%	SD: 3.4%	SD: 1.2%	SD: 7.2%	SD: 3.1%	SD: 3.0%

SD: Standard Deviation.

weights are the hyperparameters that lead to the highest accuracy at 92.7±3.4%. Although this methodology led us to the optimal hyperparameters for this dataset, our work is not yet finished. We now need to validate this model on new data that was not involved in the hyperparameter optimization process.

5.7 MODEL VALIDATION AND SELECTION

We have found the best hyperparameters for our model; however, we cannot use the accuracy score measured in the grid search cross-validation to report the accuracy of our model. It may seem counterintuitive, but let's take a step back to see why we were doing cross-validation in the first place. We wanted to determine which hyperparameters were the best to use. However, in doing so, each of the five folds had an opportunity to be the validation set, and it was for this validation set that we chose the best hyperparameters. In other words, the entire dataset was used to decide which hyperparameters were the best choice. Returning to our example shown in Tables 5.1 and 5.2, it would be ideal if we had a sixth fold of another 200 samples, consisting of 180 healthy patients and 20 sick patients. We would then train our *k*-nearest neighbors classifier with *k*=5 and distance-based weights on folds 1, 2, 3, 4, and 5. The model would then be tested on fold 6, and this metric would give us an unbiased measure of how well our model generalizes to new data. This process can be taken a step further. In the next iteration, the grid search cross-validation can be performed on folds 1, 2, 3, 4, and 6 to determine the best hyperparameters. The model will then be trained on folds 1, 2, 3, 4, and 6 using those best hyperparameters and tested on fold 5. This will be repeated with each fold having a chance to be the outer test fold. This process is called nested cross-validation and is a method to efficiently use your data to obtain a more robust metric of the generalized performance of your model. Once again, the choice of using six folds here was arbitrary.

Data leakage is another phenomenon that must be avoided when validating a model. Specifically, it is the leakage of any information used to train and optimize a model, which is also used to test the performance of this model. Some common mistakes made by first-time ML practitioners include (but are not limited to) determining the mean and standard deviation for standardization or the PCA loadings using all the samples in the dataset and not just the training samples; using samples that were involved in the hyperparameter optimization process to test model performance; and performing data augmentation (e.g., linear interpolation to generate new samples) on the entire dataset instead of just the training samples.

In the case of classification models, the use of metrics such as accuracy alone can be misleading. For instance, consider the case of a cancer test that has a very high accuracy. The doctor simply tells every patient that they do not have cancer. This will likely lead to a very high accuracy since the vast majority of people do not have cancer at a given point in time. However, the test is clearly dangerous as it will lead to many false negatives. In these instances, more balanced metrics, such as Matthews' correlation coefficient,[8] can be used, which considers the number of true positives, true negatives, false positives, and false negatives, to minimize bias obtained from using an unbalanced dataset (i.e., a dataset with an uneven number of samples in each category).

5.8 MODEL DEPLOYMENT

5.8.1 Strategies for Mitigating Uncertainties in Deploying ML Models

When using ML models in production, it is important to be aware of their limitations. Depending on the model used, even minor extrapolations outside the data used in the training set can result in very large errors. For example, a model that was trained on bioreactor data in a temperature range of 36–38°C and a pH of 7–8 should not be used when the bioreactor is operating at 34°C at pH 6.5 without first verifying the validity of its results. If a dashboard is developed to provide real-time monitoring predictions from sensor data (e.g., viable cell density from capacitance and other probes), then an alarm or indicator can be made visible when the capacitance is outside the range of the capacitance readings used to train the model. This could suggest either a malfunction in the probe or a major deviation from normal in the bioreactor.

5.8.2 Model Lifecycle Management

As discussed in the previous section, it is important to stay within the range defined by the training data used to train a model when using it in production. However, one aspect that is difficult to quantify is how the instrumentation itself may drift over time. Motors used for stirring can slow down or lose torque over time. Thermocouples and pH probes can lose their calibration. The surface within the bioreactor may become rough over time, impacting the adhesion of cells to the bioreactor walls. There are several strategies that can be deployed to ensure a model is well-maintained. For instance, one strategy can be to retrain the model every week using only data obtained in the last six months (for example). This ensures that the model is always reflective of the state of the bioreactor at present. Another strategy can be to perform regular quality control of real-time monitoring predictions by comparing it to offline measurements. For example, a prediction made by a glucose monitor can be verified using an offline enzymatic assay. If the error drifts outside an acceptable range, the model may be due for retraining on fresh data.

REFERENCES

1. Hoerl, A. E.; Kennard, R. W. Ridge Regression: Biased Estimation for Nonorthogonal Problems. *Technometrics* **1970**, *12* (1), 55–67. DOI: 10.1080/00401706.1970.10488634.
2. Wold, S.; Ruhe, A.; Wold, H.; Dunn, I., W. J. The Collinearity Problem in Linear Regression. The Partial Least Squares (PLS) Approach to Generalized Inverses. *SIAM Journal on Scientific and Statistical Computing* **1984**, *5* (3), 735–743. DOI: 10.1137/0905052.
3. Fix, E.; Hodges, J. L. Discriminatory Analysis. Nonparametric Discrimination: Consistency Properties. *International Statistical Review/Revue Internationale de Statistique* **1989**, *57* (3), 238–247. DOI: 10.2307/1403797 (accessed 2025/12/16/). JSTOR.
4. Lozupone, C.; Knight, R. UniFrac: A New Phylogenetic Method for Comparing Microbial Communities. *Applied and Environmental Microbiology* **2005**, *71* (12), 8228–8235. DOI: 10.1128/AEM.71.12.8228-8235.2005.
5. Shrivastava, R.; Mahalingam, H.; Dutta, N. N. Application and Evaluation of Random Forest Classifier Technique for Fault Detection in Bioreactor Operation. *Chemical Engineering Communications* **2017**, *204* (5), 591–598. DOI: 10.1080/00986445.2017.1292259.
6. Jin, Y.; Qin, S. J.; Huang, Q.; Saucedo, V.; Li, Z.; Meier, A.; Kundu, S.; Lehr, B.; Charaniya, S. Classification and Diagnosis of Bioprocess Cell Growth Productions Using Early-Stage Data. *Industrial & Engineering Chemistry Research* **2019**, *58* (30), 13469–13480. DOI: 10.1021/acs.iecr.9b01175.
7. Zhang, H. The Optimality of Naive Bayes. *American Association for Artificial Intelligence* **2004**, *1* (2), 3.
8. Matthews, B. W. Comparison of the Predicted and Observed Secondary Structure of T4 Phage Lysozyme. *Biochimica et Biophysica Acta (BBA) – Protein Structure* **1975**, *405* (2), 442–451. DOI: 10.1016/0005-2795(75)90109-9.

6 PAT and Process Control

Xingge Xu, Rupa Haldavnekar, and Malvika Nagrath

6.1 PAT FRAMEWORK FOR CONTROL STRATEGY

Process Analytical Technology (PAT) was introduced in biomanufacturing to enhance the efficiency and robustness of biologics production [1, 2]. The implementation of PAT in biologics manufacturing optimizes resource utilization while enabling the consistent production of high-quality products with enhanced titers [2]. PAT provides a systematic framework for understanding and controlling bioprocesses by integrating analytical tools, data-driven models, and control strategies [1, 2]. The implementation of PAT involves several interconnected elements that typically follow a structured sequence:

1. ***Identification of Measurable Attributes***: Determine the critical process parameters (CPPs) and that influence product quality and performance.
2. ***Selection of Process Analyzers and Measurement Systems***: Choose appropriate analytical technologies and sensors capable of accurately monitoring the identified attributes in real time.
3. ***Data Acquisition***: Collect and record data on the measurable attributes using the selected analyzers and measurement systems.
4. ***Process Modeling and Control Strategy Development***: Build predictive models based on the acquired data to describe process behavior and define suitable control modes.
5. ***Process Control Implementation***: Apply the developed control strategies to maintain CPPs within predefined limits, ensuring that critical quality attributes (CQAs) remain within the established design space.

The core objective of PAT is to foster deep process understanding through continuous measurement, analysis, and monitoring of CPPs to develop a strategy for production process control. Once process understanding is achieved, the process control framework can actively regulate CPPs to maintain CQAs aligned with regulatory requirements. This chapter explores the various elements, variables, and methodologies that underpin PAT-based process control in biomanufacturing.

6.2 ELEMENTS AND VARIABLES OF PROCESS CONTROL

In any manufacturing process, production begins with raw materials that undergo a series of established procedures that yield the final product. Controls can be implemented at multiple stages—starting from raw material acquisition to the execution of these procedures—to ensure the desired product quality [1]. Similarly, in biologics

DOI: 10.1201/9781032662961-6

manufacturing, control strategies are applied at critical points within the production workflow to achieve high-quality biologics with optimal yields.

Within the PAT framework, measurement, analysis, and monitoring of CPPs to ensure CQAs are accomplished using advanced analytical tools equipped with sensors and supported by multivariate data analysis, since biologic manufacturing is a complex process [1, 2]. Consequently, process control strategies integrate these elements to maintain process stability and product consistency [1, 2].

Process control can be implemented manually, e.g., adjusting a thermostat to regulate home temperature. However, the requirements of continuous manufacturing in modern bioprocessing necessitate automation to ensure precision, consistency, and efficiency [1]. Automation in bioprocessing is built upon understanding the *cause-and-effect relationships* among three key types of variables: **process variables**, **control (or manipulated) variables**, and **disturbance variables** [3].

- ***Process Variable (Controlled or Measured Variable)***: This is the output of a dynamic process, such as a cell culture that can be measured and monitored. Common examples include dissolved oxygen (DO) and pH levels within the bioreactor. Each process variable typically has a predefined **set point** that represents the desired operating condition [3].
- ***Control Variable (Manipulated Variable)***: This is an input variable that exerts a detectable influence on the process. By adjusting the control variable, the system can regulate the process variable to maintain it at its set point [3]. For instance, adding acid or alkali to the culture system is a control action that adjusts pH, where pH is the process variable, and the addition of acid or alkali is the control variable.
- ***Disturbance Variable***: Disturbances are unintended factors that affect process variables, often originating from external conditions [3]. For example, cell metabolism can produce lactate as a by-product, indirectly impacting pH control in the bioreactor.

As discussed in the previous section on process monitoring, glucose concentration in cell culture serves as a key CPP influencing multiple CQAs, including glycosylation patterns that directly affect productivity and product quality [4]. To maintain glucose within an optimal range, analytical tools such as Raman spectroscopy and in-line turbidity sensors can be employed [2, 5]. These tools enable real-time assessment of glucose levels and facilitate automated control of glucose feeding, thereby sustaining an ambient cellular environment conducive to high-yield production [5].

To illustrate the concept of process control (Figure 6.1), consider the example of glucose concentration management. The process variable, glucose concentration, is measured using sensors (e.g., Raman spectroscopy or turbidity sensors). A predefined set point (11 mM in this example) is established, and the controller calculates the deviation between the measured value and the set point (known as the **set-point error**). Based on this error, the controller determines the corrective action, which is executed by an actuator—such as adjusting the glucose feed rate (controlled variable)—to restore the process to its desired state. Outside of this example, other

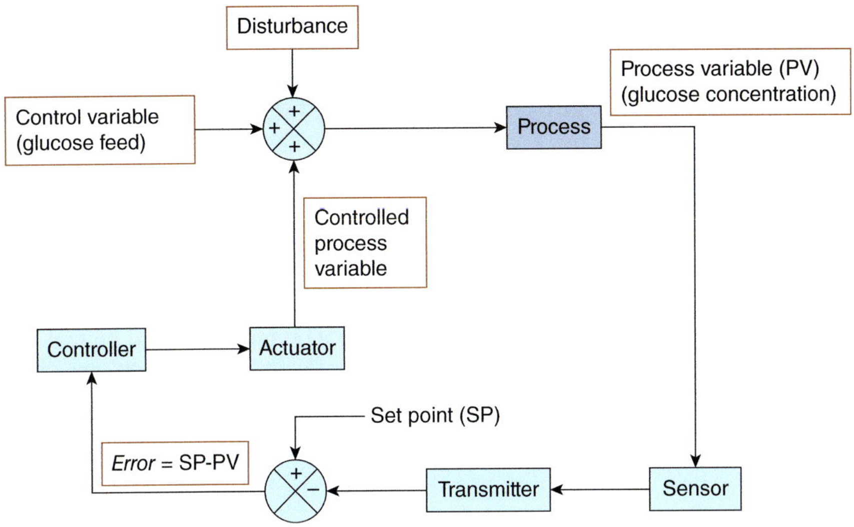

FIGURE 6.1 Process control schematic.

manipulated variables (such as temperature, pH, agitation rate, and so on) can also receive corrective actions from the respective actuator as needed. It must be kept in mind that the control loop is unique to a process, and its development is based on the process understanding acquired by the PAT framework.

Effective control should be rapid, accurate, and resilient to disturbances [1]. Controllers rely on mathematical algorithms to compute and implement corrective actions [1], which will be explored in detail in a subsequent section of this chapter.

6.3 TYPES OF PROCESSES AND CONTROL STRATEGIES IN PAT FRAMEWORKS

In PAT-enabled manufacturing, the classification of process types is inseparable from the choice of control strategy. The time dependence, degree of nonlinearity, and variability inherent to batch, fed-batch, and continuous operations dictate whether simple feedback, classical control, or adaptive approaches are appropriate or required.

6.3.1 Batch and Fed-Batch Processes

Batch and fed-batch processes dominate upstream biologics manufacturing and are characterized by strong time-varying dynamics, non-stationarity, and phase-dependent behavior. Early-stage process control in these systems often relies on on/off (discrete) control, particularly for safety-critical variables such as temperature limits, foam control, or level regulation [6]. While robust and simple, on/off control is inherently reactive and may introduce oscillations, making it unsuitable for tight quality control.

For most regulatory and operational variables—such as pH, DO, agitation speed, and temperature—PID control remains the industrial standard [7]. PID controllers provide continuous corrective action and are well-suited for maintaining local set points during relatively stable phases of batch or fed-batch operation. However, fixed PID parameters are typically tuned for a specific operating region and may perform poorly as cell density, viscosity, and metabolic activity evolve over time.

To address this limitation, PAT enables the deployment of adaptive control strategies, where controller parameters or control laws are updated in response to changing process dynamics [8]. For example, adaptive PID tuning informed by real-time spectroscopic PAT measurements can compensate for shifts in oxygen uptake rate or metabolic state during fed-batch culture. In this context, PAT-derived soft sensors play a critical role by providing timely estimates of unmeasured states that govern system dynamics.

6.3.2 Continuous and Hybrid Processes: Control Implications

Continuous bioprocesses, by design, aim to operate near steady state and are more amenable to classical control approaches [9]. PID control is highly effective for maintaining flow rates, pressures, and concentrations under nominal conditions, while on/off control is often retained as a protective layer for alarms and interlocks.

However, biological systems rarely remain truly stationary over long durations. Slow drift due to cell line evolution, fouling, or raw material variability can degrade PID performance [10]. In such cases, adaptive control can be useful to maintain long-term stability and product quality. PAT measurements enable continuous reassessment of process behavior, allowing controllers to adjust gains, update model parameters, or switch control modes as needed.

Hybrid process configurations—combining batch upstream operations with continuous downstream purification or viceversa—require a hierarchical control structure in which different control strategies coexist. Fast, local variables are typically regulated by PID loops, while adaptive supervisory control manages interactions between discrete and continuous unit operations using PAT data streams.

6.4 PROCESS DYNAMICS AND CONTROL STRATEGY SELECTION

6.4.1 Dynamic Response and Time Delays

Bioprocesses often exhibit long time constants and measurement delays, particularly for CQAs that depend on cellular responses [11]. On/off control is ill-suited for such dynamics due to excessive oscillations. PID control, while effective for fast variables, may struggle with delayed responses unless augmented by feedforward or model-based elements.

PAT mitigates these challenges by providing high-frequency, indirect measurements (e.g., Raman, NIR), enabling adaptive and predictive control strategies that anticipate system response rather than reacting to delayed signals.

6.4.2 Nonlinearity and Operating Regime Changes

Strong nonlinearity and variable coupling limit the effectiveness of fixed-gain PID controllers across an entire batch or campaign. Adaptive control, including gain scheduling or model-updating approaches, allows control laws to evolve as the process transitions between growth phases or operating regimes. PAT data provide the real-time context needed to identify these transitions reliably [12].

6.4.3 Disturbances and Drift

Disturbances may be abrupt (e.g., feed interruptions) or gradual (e.g., metabolic drift). On/off control provides a fail-safe response to extreme deviations, PID control suppresses routine disturbances, and adaptive control addresses systematic performance degradation over time [6]. Together, these layers can be used to form a robust control architecture supported by PAT analytics.

The dynamic, nonlinear, and time-varying nature of bioprocesses necessitates a layered control architecture in which multiple control strategies operate in a coordinated manner [13]. At the lowest level, regulatory control is typically implemented using on/off and PID controllers to maintain fast, safety-critical, and operational variables such as temperature, pH, DO, and pressure within predefined limits. On/off control provides robust fail-safe protection and alarm-based intervention, while PID control enables continuous set-point tracking and disturbance rejection under nominal operating conditions. At a higher level, supervisory control manages slower, quality-relevant variables by adjusting set-points, feed profiles, or operating modes based on process phase and trajectory information derived from PAT measurements. Building on this foundation, predictive and adaptive control strategies leverage real-time PAT data, soft sensors, and process models to account for nonlinearity, biological variability, and drift, dynamically updating control actions as process dynamics evolve. Together, these layers enable stable operation, improved product quality, and increased process robustness across batch, fed-batch, and continuous biomanufacturing systems.

6.5 BASIC CONTROLLER ACTIONS

The purpose of the controllers is to provide automation to the process and automatic controller actions can either be discontinuous (on/off control), continuous (PID control), or advanced (adaptive control) [1, 3].

6.5.1 On/Off Control

On/off control is generally employed for the PVs that change slowly, such as temperature and pressure. It is the oldest and least expensive type of controller because the actuators are simpler and cheaper than the ones used in continuous controllers. This type of controller is usually used for temperature controls in bioreactors [3].

6.5.2 Proportional–Integral–Derivative (PID) Control

PID control is a widely used continuous control system that integrates feedback mechanisms. These systems are valued for their simple structure, ease of implementation, and low maintenance requirements, making them among the most cost-effective solutions for process control applications [1, 14]. Due to their simplicity and versatility, PID controllers are employed across diverse domains, including self-driving vehicles, autonomous robotics, and pharmaceutical and biologics manufacturing [1, 14].

As the name implies, proportional–integral–derivative (PID) control comprises three components: proportional (P), integral (I), and derivative (D).

- The **proportional term** addresses the instantaneous error between the set point and the process variable to achieve the desired output. Proportional algorithm is given by: $u(t) = K_p e(t)$, where K_p is the proportional gain and $e(t)$ is the error between the set point and process variable [14–16].
- The **integral term** accounts for the accumulated error over time, helping eliminate steady-state offsets and ensuring long-term accuracy. Integral algorithm is given by: $u(t) = K_i \int_0^t e(t)\,dt$, where K_i is the integral gain [14–16].
- The **derivative term** responds to the rate of change of the error, providing predictive control to enhance control performance. A derivative algorithm is given by: $u(t) = K_d \frac{d}{dt} e(t)$, where K_d is the derivative gain [14–16].

Together, these components enable dynamic supervision of processes and corrective adjustments to maintain optimal performance. Depending on application requirements, control strategies may employ any combination of these terms—such as P, PI, PD, or full PID control.

6.5.2.1 Structure of PID Control

The three control functions in a PID control system can be organized in different configurations; however, the **parallel** and **series** structures are the most employed [14].

The **parallel control structure** is typically expressed using the following algorithm: $(t) = K_c\left(e(t) + \frac{1}{T_i}\int e(t)\,dt + T_d \frac{d}{dt} e(t)\right)$, where the proportional gain, $K_p = K_c$, process gain; T_i is the integral time given by $K_i = \frac{K_c}{T_i}$; T_d is the derivative time given by $K_d = K_c T_d$.

In the **parallel configuration**, the proportional (P), integral (I), and derivative (D) functions operate independently within separate algorithms, and their outputs are combined by summation. In this structure, the process gain, K_c, influences all three control components—P, I, and D—ensuring a unified scaling effect across the control actions [14].

The **series control structure** is represented by the algorithm below:

$$e_1(t) = e(t) + T_d \frac{de(t)}{dt} \text{ and } u(t) = K_c\left(e_1(t) + \frac{1}{T_i}\int e(t)\,dt\right)$$

In the **series configuration**, the proportional (P), integral (I), and derivative (D) functions interact with each other, allowing the integral and derivative actions to influence the proportional component. This interaction creates a cascaded effect within the control loop. Similar to the parallel structure, the process gain, K_c, applies to all three control functions—P, I, and D—ensuring consistent scaling across the system [14].

6.5.3 Adaptive Controllers

Biologics manufacturing is the result of a bioprocess that is inherently complex and nonlinear. Controllers for autonomous, continuous biomanufacturing must address intricate multivariate challenges through the use of analytical models and intelligent control systems [1, 17]. These models and systems help minimize variability in CPPs and CQAs, enabling a leaner and more cost-effective manufacturing process [17]. Once a comprehensive understanding of the process is established, adaptive control strategies can facilitate real-time release testing, effectively managing interactions among process variables, dynamics, and disturbances [1].

Due to the complexity and nonlinearity of bioprocesses, certain process variables may remain unknown, unmeasurable, or subject to temporal changes—factors that can compromise the performance of control systems. Adaptive controllers are designed to automatically adjust the control system to achieve the desired outcomes [17, 18]. Similar to conventional feedback-based control systems, adaptive controllers operate by establishing a performance index. The deviation between the desired and actual performance index is fed back into the adaptive controller, which then modifies the unmeasured, unknown, or changing process variables. These adjustments ultimately enhance the overall performance of the process control system [18].

6.6 BASIC CONTROL SCHEMES IN PAT-ENABLED BIOPROCESSING

The increasing complexity of modern biomanufacturing has driven the evolution of control strategies from simple, recipe-based approaches to advanced, model-supported frameworks [19]. Within the PAT paradigm, control schemes are distinguished by how they utilize measurements, feedback, and process knowledge to manage variability, disturbances, and dynamic behavior. Open-loop, closed-loop, inferential, and model-supported predictive control represent successive levels of sophistication, each playing a distinct role in contemporary control architectures.

6.6.1 Open-Loop Control

Open-loop control operates without direct feedback from the process output, relying instead on predefined control actions or time-based recipes. In bioprocessing, open-loop strategies are frequently used for operations such as programmed feed additions, temperature shifts, or induction of protein expression at scheduled times [20]. These approaches are straightforward to implement and computationally inexpensive, making them attractive for early-stage development and routine manufacturing. However, open-loop control assumes that process behavior is repeatable and predictable, an assumption that is often not applicable to biological systems due to raw material variability, stochastic cell behavior, and environmental disturbances. Within PAT-enabled environments, open-loop control is increasingly complemented by real-time analytics, which allow deviations from expected trajectories to be detected even if corrective action is not automatically applied.

6.6.2 Closed-Loop Control

Closed-loop control introduces feedback by continuously measuring process variables and adjusting inputs to minimize deviations from desired set points. PID controllers form the backbone of closed-loop control in biomanufacturing, regulating key operational parameters such as pH, DO, temperature, agitation speed, and pressure [21]. The effectiveness of closed-loop control depends strongly on measurement frequency, signal quality, and controller tuning. PAT enhances closed-loop performance by providing rapid, noninvasive measurements—such as spectroscopic signals—that reduce effective time delays and improve disturbance rejection. Nevertheless, classical closed-loop controllers typically rely on fixed parameters and may struggle when process dynamics change significantly across operating regimes or process phases.

6.6.3 Inferential Control

Inferential control extends the closed-loop paradigm by enabling feedback on variables that cannot be measured directly or in real time. In bioprocesses, many CQAs, including biomass concentration, metabolite levels, and product quality metrics, are either measured offline or not immediately observable. Inferential control addresses this limitation by employing soft sensors that infer these variables from readily available online measurements, often using multivariate statistical models or machine learning [22, 23]. PAT plays a central role in inferential control by supplying high-dimensional data streams—such as Raman, near-infrared, or mass spectrometry signals—that serve as proxies for underlying biological states. The inferred variables can then be incorporated into closed-loop control, enabling more direct regulation of product quality and process performance.

6.6.4 Model-Supported Predictive Control

Model-supported predictive control represents the most advanced class of basic control schemes, leveraging dynamic process models to anticipate future behavior and optimize control actions over a defined prediction horizon. Unlike conventional feedback control, predictive control explicitly accounts for multivariable interactions, process constraints, and time delays. In PAT-enabled bioprocessing, these models are often hybrid in nature, combining mechanistic representations of biological kinetics with data-driven components calibrated using real-time analytical measurements. Model-supported predictive control allows control decisions to be evaluated before implementation, enabling proactive responses to disturbances and gradual drift [24]. While widespread industrial adoption in biologics manufacturing remains limited due to model maintenance and regulatory considerations, this approach forms the conceptual foundation for emerging digital twin–based control strategies and adaptive, self-optimizing manufacturing systems.

6.6.5 Regulatory Perspective Summary

Collectively, these control schemes reflect a regulatory continuum rather than mutually exclusive options [25]. FDA PAT guidance encourages manufacturers to progressively integrate feedback, inference, and prediction into their control strategies, using real-time data to demonstrate process understanding and ensure consistent quality. In this context, PAT-enabled control schemes serve not only as operational tools but also as regulatory enablers for flexible, robust, and knowledge-driven biomanufacturing.

6.7 APPLICATIONS OF PAT FOR PROCESS CONTROL

Process control is a cornerstone of the PAT framework for continuous manufacturing [1]. It enables real-time monitoring and regulation of CPPs to ensure that CQAs are consistently met, thereby supporting the production of high-quality biologics. These control strategies can be applied across different manufacturing stages, upstream, downstream, and formulation, where CPPs have been identified. Automated and advanced control systems play a vital role in maintaining operations within the defined design space. While multiple control systems can be implemented to achieve desired outcomes, it is essential to emphasize the importance of selecting appropriate process variables for effective control.

Kornecki and Strube [2] exemplified this principle by cultivating Chinese hamster ovary (CHO) cells in a bioreactor for monoclonal antibody (mAb) production. They employed in-line turbidity measurements alongside offline Raman spectroscopy coupled with chemometric analysis (PLS) to predict viable cell concentration. Turbidity probes estimate total cell concentration based on optical density, which is often assumed to represent viable cell concentration. In contrast, Raman spectroscopy combined with PLS can predict substrate concentrations and viable cell concentrations offline. Their findings revealed that Raman-based predictions successfully detected declining viable cell concentrations after 234 hours, whereas turbidity probes failed to do so. The discrepancy arose because cell accumulation around the turbidity probe led to erroneous readings, suggesting an increase in viable cell concentration. Adjusting the probe reduced this error, but the issue underscores the criticality of selecting the right process variable.

In this example, if viable cell concentration closely correlated with total cell concentration, turbidity data could serve as a suitable process variable for control. However, in the case of mAb production from CHO cells, this assumption did not hold true, making Raman data the more reliable choice. This case study highlights that process control effectiveness depends not only on the sophistication of the control system but also on the appropriateness of the chosen process variable.

6.8 CONCLUSIONS

The integration of PAT into bioprocess control represents a paradigm shift toward data-driven and adaptive manufacturing. By enabling real-time measurement of process variables, PAT allows deviations of process variables from set points to be detected and translated into appropriate control actions according to the selected control strategy. Traditional on/off and PID control schemes can be ill-suited to bioprocesses due to the inherent nonlinear behavior, time delays, and time-varying dynamics of the latter. In contrast, model-based and adaptive control strategies are better equipped to accommodate such complexities. As control architectures progress from open-loop to closed-loop, inferential, and ultimately model-supported predictive control, the dependence on PAT increases correspondingly. Advanced model predictive control strategies rely on PAT not only to provide real-time process measurements but also to enable state estimation, model updating, and prediction of future process trajectories. The selection and deployment of appropriate PAT are therefore critical to achieving reliable control performance and optimizing process outcomes, such as yield and product quality. Together, PAT and process control establish bidirectional information exchange between the physical process and its virtual representation, thereby fulfilling a key functional requirement of digital twins. In this context, PAT-enabled process control constitutes a foundational step toward autonomous and intelligent biomanufacturing, aligned with the principles of Biopharma 4.0.

REFERENCES

1. Schlindwein, W. S., & Gibson, M. (Eds.), *Pharmaceutical Quality by Design – A Practical Approach*, in Advances in Pharmaceutical Technology, 2018, John Wiley & Sons Ltd.
2. Kornecki, M. and J. Strube, *Process analytical technology for advanced process control in biologics manufacturing with the aid of macroscopic kinetic modeling.* Bioengineering (Basel), 2018. **5**(1).
3. Ryniecki, A., *Basics of process control: The on-off control system.* PrzemysŁ SpoŻywczy, 2015. **1**(11): p. 28–31.
4. Kozma, B., A. Salgo, and S. Gergely, *On-line glucose monitoring by near infrared spectroscopy during the scale up steps of mammalian cell cultivation process development.* Bioprocess Biosyst Eng, 2019. **42**(6): p. 921–932.
5. Domjan, J., et al., *Raman-based dynamic feeding strategies using real-time glucose concentration monitoring system during adalimumab producing CHO cell cultivation.* Biotechnol Prog, 2020. **36**(6): p. e3052.
6. Rathore, A.S., et al., *Bioprocess control: Current progress and future perspectives.* Life (Basel), 2021. **11**(6).
7. Jones, J., et al., *Improved control strategies for the environment within cell culture bioreactors.* Food and Bioprod Process, 2023. **138**: p. 209–220.
8. Williams, T., et al., *Machine learning and metabolic modelling assisted implementation of a novel process analytical technology in cell and gene therapy manufacturing.* Sci Rep, 2023. **13**(1): p. 834.
9. Pappenreiter, M., et al., *Model predictive control for steady-state performance in integrated continuous bioprocesses.* Bioprocess Biosyst Eng, 2022. **45**(9): p. 1499–1513.

10. Pugh, P.C., et al., *Predicting raw material impact on cell culture parameters in commercial biotherapeutic manufacturing.* Biochem Biophys Rep, 2025. **43**: p. 102192.
11. Ajit Gopalakrishnan, N.S.K., and S. Narasimhan, *Incorporating delayed and infrequent measurements in Extended Kalman Filter based nonlinear state estimation.* J Process Control, 2011. **21**: p. 119–129.
12. Craven, S., J. Whelan, and B. Glennon, *Glucose concentration control of a fed-batch mammalian cell bioprocess using a nonlinear model predictive controller.* J Process Control, 2014. **24**(4): p. 344–357.
13. Goulart, D.A., R.D. Pereira, and F.V. Silva, *AutoRL framework for bioprocess control: Optimizing reward function, architecture, and hyperparameters.* Digit Chem Eng, 2025. **16**.
14. Borase, R.P., et al., *A review of PID control, tuning methods and applications.* Int J Dyn Control, 2020. **9**(2): p. 818–827.
15. Dubey, V.G., H. Sharma, and C. Prakash, *Data Engineering for Smart Systems,* Proceedings of *SSIC.* 2021, Springer.
16. Li, Y.A., K. Heong, and G.C.Y. Chong, *PID control system analysis and design.* IEEE Contr Syst Mag, 2006.
17. Lupu, C., and M.H. Dumitru Popescu, *Adaptive Control of Bioprocess,* in Proceedings of the 10th Mediterranean Conference on Control and Automation - MED2002 Lisbon, Portugal, July 9-12, 2002.
18. Landau, I.D., *Controls, Adaptive Systems,* in Encyclopedia of Physical Science and Technology. 2003, Academic Press. p. 649–658.
19. Drobnjakovic, M., et al., *Current challenges and recent advances on the path towards continuous biomanufacturing.* Biotechnol Prog, 2023. **39**(6): p. e3378.
20. Gasset, A., et al., *Innovative bioprocess strategies combining physiological control and strain engineering of pichia pastoris to improve recombinant protein production.* Front Bioeng Biotechnol, 2022. **10**: p. 818434.
21. Chevalier, M., et al., *Design and analysis of a proportional-integral-derivative controller with biological molecules.* Cell Syst, 2019. **9**(4): p. 338–353.e10.
22. Brunner, V., et al., *Challenges in the development of soft sensors for bioprocesses: A critical review.* Front Bioeng Biotechnol, 2021. **9**: p. 722202.
23. Rathore, A.S., S. Nikita, and N.G. Jesubalan, *Digitization in bioprocessing: The role of soft sensors in monitoring and control of downstream processing for production of biotherapeutic products.* Biosens Bioelectrons: X, 2022 **12**.
24. Mu'azzam, K., et al., *A roadmap for model-based bioprocess development..* Biotechnol Adv, 2024. **73**: p. 108378.
25. Guidance for industry PAT – A framework for innovative pharmaceutical development, manufacturing, and quality assurance – U.S. Department of Health and Human Services, Food and Drug Administration, Center for Drug Evaluation and Research (CDER), Center for Veterinary Medicine (CVM), Office of Regulatory Affairs (ORA), Pharmaceutical CGMPs, September 2004.

7 PAT for Process Modelling

Xingge Xu and Shawninder Chahal

7.1 INTRODUCTION

Process modelling provides a mathematical representation of the relationships between process inputs and outputs. In biopharmaceutical manufacturing, these relationships are inherently complex due to the biological nature of the products and the coupling of physical, chemical, and biological phenomena across multiple length and time scales, ranging from nanoscale molecular and genetic engineering to macroscale process geometry and operating conditions. Process analytical technologies (PATs) are therefore essential for process modelling as they enable the acquisition of quantitative measurements needed to parameterize, calibrate, and validate the bioprocess models.

The origins of bioprocess modelling date back to the early 1900s, when microbial growth kinetics were first described [1]. Over the past century, advances in analytical assays and sensing technologies have expanded both the subjects and scope of process modelling. More recently, the emergence of real-time monitoring technologies, together with substantial increases in computational capability, has transformed process modelling from a descriptive tool into a critical enabler of process control and optimization. Today, bioprocess models are applied across several key areas, as described in the following sections [2].

7.1.1 Enhancing Process Understanding

Process models help researchers understand the underlying mechanisms of bioprocesses by allowing hypotheses to be tested under different modelling assumptions, such as first- or second-order reaction kinetics. By introducing model parameters that quantitatively describe biological states and processes, models enable systematic tracking of dynamic changes during bioprocess operation. For example, solving a metabolic flux model to estimate intracellular metabolite conversion rates provides insight into metabolic shifts occurring during cell culture.

7.1.2 Scale-Up and Scale-Down

Bioprocesses typically begin at laboratory scale before being scaled up for production. Scaling up is not a simple matter of multiplying physical dimensions; latent parameters such as oxygen transfer efficiency (kLA) must be maintained. Modelling kLA as a function of bioreactor geometry, sparger and impeller design,

DOI: 10.1201/9781032662961-7

and operating conditions is essential for successful scale-up. Conversely, scale-down models ensure that validation studies at smaller scales accurately represent large-scale operations.

7.1.3 In-Silico Design of Experiment/Process Optimization

Simulation-based design of experiments (DOE) and optimization reduce experimental costs and help identify optimal operating conditions. These simulations also support model predictive control strategies that leverage historical and real-time data for improved process outcomes.

7.1.4 Soft Sensors

Certain critical process parameters (CPPs) or critical quality attributes (CQAs) cannot be measured directly. Models bridge this gap by linking measurable variables to latent variables using mechanistic relationships or machine learning algorithms, enabling indirect monitoring. This concept aligns with soft sensors, as discussed in Chapter 3.

7.1.5 Economic Evaluation and Life-Cycle Management

By simulating material and equipment usage under varying conditions, process models facilitate economic assessments and life-cycle management without extensive physical experimentation.

7.2 MODELLING FRAMEWORK

7.2.1 Model Subject

Depending on the modelling objective, the subject of the model can encompass biological and biochemical variables, physicochemical parameters, system control settings, and equipment characteristics. A basic bioprocess model describes interactions among biological and biochemical variables—such as biomass, metabolites, nucleic acids, proteins, and viral particles. These variables can be classified as CQAs or CPPs.

Beyond biological and biochemical factors, physicochemical conditions (e.g., dissolved oxygen, temperature, viscosity, pH) and system control outputs (e.g., flow rate, feed rate, and stirring rate) are CPPs that strongly influence process dynamics. Behind the control outputs, process control configuration, including the valves, pumps, and algorithms for executing control strategies (e.g., proportional integral derivative [PID]), can be considered. Lastly, material and geometric factors (e.g., reactor dimensions, impeller design, sparger configuration, filter material) affect heat and mass transfer during the process. Modelling these aspects ensures scalability and transferability across different setups.

7.2.2 Modelling Scopes

Bioprocesses typically comprise three consecutive sections: upstream, where biological products are generated; downstream, where products are purified; and formulation, where products are prepared for delivery. Each section can include several sequential unit operations. Although these steps are often modelled individually, they remain interconnected because the output of one step commonly serves as the input to the next.

***Upstream*:** Models describe concentrations of biological and biochemical CQAs and CPPs and their production rates or consumption rates. These models are essential for understanding process kinetics and optimizing material ratios (e.g., multiplicity of infection, medium composition). When physicochemical factors and operational controls are included, the model gains capability for optimizing operating conditions and control strategies.

***Downstream*:** Biological activity ceases, and the focus shifts to isolating the product from impurities through mechanical operations. Models here emphasize recovery and impurity removal, where the interactions with physicochemical factors and operational controls are mostly considered.

***Formulation*:** The primary biological variable is the product itself. Liquid formulations focus on stability and efficiency influenced by excipients, while solid formulations require an additional lyophilization step, modelled using thermodynamic principles to describe freeze-drying kinetics. For messenger RNA-Lipid Nanoparticle (mRNA–LNP) formulations, encapsulation rates and particle size of mRNA by lipid nanoparticles are considered.

7.2.3 PAT for Modelling Data

Model scope is inherently constrained by data availability. PATs play a pivotal role by providing both input data and validation information. Therefore, the reliability and capability of any process model depend heavily on the accuracy and timelines of the PATs.

As models incorporate more variables with inherent measurement uncertainties, prediction uncertainty also increases. Even small variations in input data can lead to significant deviations in outputs beyond the tolerance limit of biomanufacturing, which demand robust monitoring and control to ensure product quality and reliability.

As discussed in Chapter 3, a trade-off trend exists between timelines and accuracy for the off-line, at-line, in-line, and on-line PATs [3]. While real-time or near-real-time PATs have been developed for many biological variables, more variables still require off-line analysis. On the other hand, physicochemical conditions and system control outputs are usually recorded in real time by sensors and actuators. When only off-line measurements are available, models can support functions such as process understanding, scale-up and scale-down studies, model-assisted DOE, and economic evaluations. In contrast, real-time functionalities—such as soft sensors, real-time optimization, and model-based predictive control—require continuous PAT data streams.

7.3 TYPE OF MODELS

7.3.1 Mechanistic Models

7.3.1.1 Unstructured Cell Growth Modelling

Mechanistic models are based on first principles of physical, chemical, or biological laws. A basic example is cell duplication under ideal conditions, assuming a constant growth rate without nutrient or waste limitations:

$$\frac{dX}{dt} = \mu X \text{ or } X = X_0 e^{\mu t} \tag{7.1}$$

where X is the cell density, μ is the specific cell growth rate, and X_0 is the initial condition of the cell density. This exponential growth model applies when nutrients are abundant. However, growth rate is limited by nutrient availability and byproduct accumulation. To account for these limitations, Monod adapted the Michaelis–Menten enzymatic kinetics to describe how specific growth rate changes with substrate concentration [4]:

$$\frac{dX}{dt} = \mu_{max} \frac{S}{s + k_s} X \tag{7.2}$$

where μ_{max} is the maximum specific growth rate, S is the substrate concentration (e.g., glucose or glutamine), and k_s is half saturation concentration of the substrate. This relationship describes an S-shaped growth curve that saturates when substrate is depleted.

Further refinements include the effects of inhibitors and cell death:

$$\frac{dX}{dt} = \left(\mu_{max} \frac{S}{s + k_s} \frac{k_I}{I + k_I} - kd_{max} \frac{I}{I + kd_I} \frac{kd_s}{s + kd_s} \right) X \tag{7.3}$$

where I is the growth inhibitor concentration (e.g., ammonia, lactate), k_I is the half saturation constant for inhibition, and kd_{max} is the maximum cell death rate due to nutrition limitation and byproduct accumulation. The modelling framework is named Monod kinetics. Other kinetics have also been proposed to describe the S curve cell growth because of substrate limitation and inhibition, such as Blackman [5], Teissier [6], and Moser [7].

The consumption and production of biochemical products by cells can be expressed as follows:

$$\frac{dS}{dt} = k_{S/X} X \tag{7.4}$$

where $k_{S/X}$ is the consumption rate of S by X. When the $k_{\frac{S}{X}}$ and μ are constant, Equation (7.4) can be rewritten as follows:

$$dX = Y_{X/S} dS \tag{7.5}$$

where the $Y_{X/S}$ is the yield coefficient of X from S. Both the $k_{S/X}$ and $Y_{X/S}$ can vary under different metabolic states. These coefficients during different stages of cell growth help researchers understand the metabolic shifts, and estimating them in real-time gives researchers insights to maintain the system in the most productive statues [8].

7.3.1.2 Structured Model

The models described above are unstructured, meaning they treat the cell as a single entity, even though biochemical reactions occur in various locations within the cell [9]. In contrast, structured models represent a single cell as a set of individual and interconnected compartments. One example is modelling metabolic states using metabolic flux analysis (MFA), which describes dozens of biochemical reactions within the metabolic network, such as those in glycolysis or the citric acid cycle. Example equations include [10] the following:

$$Glucose \leftrightarrow 2Pyruvate + 2ATP + 2NADH$$

$$Phenylalanine + O_2 + NADPH \ + H^+ \rightarrow Tyrosine$$

The chemical formulas and stoichiometric coefficients can be known or fitted. The flux of each reaction can be calculated by tracking changes in biochemical variables over time.

Cells are not always the ultimate production target in many bioprocesses, such as protein production or viral vector manufacturing. For protein production like monoclonal antibody (mAB) production, unstructured models often assume mAB production is proportional to the cell density [11] or energy source consumption [12] with or without coefficients influenced by metabolic state and physicochemical conditions, depending on the model complexity. In contrast, a structured model, such as a genome-scale model of Chinese Hamster Ovary (CHO) cells, captures processes from gene expression to metabolic reactions, protein synthesis, and cell proliferation [13]. Viral vector production introduces additional complexity, as it typically involves two distinct phases: cell growth and viral production, separated by transfection, infection, or transduction to introduce viral genomes into cells. An unstructured model approximates viral particle production using a time-dependent S-curve, such as a logistic function, while a structured model accounts for multiple cellular-scale reactions, including viral uptakes, viral gene replication, viral gene transcription, and viral particle assembly [14].

7.3.1.3 Influence of Physicochemical Variables, Control Outputs, and Material/Geometric Factors

The effects of physicochemical variables on cell behaviour can also be incorporated into the rate constants of cell growth models. For example,

$$\mu_{max} = \mu_{opt} \ \gamma(pH) \ \gamma(temp) \tag{7.6}$$

where μ_{opt} is the optimal specific growth rate under optimal condition, and $\gamma(pH)$ and $\gamma(temp)$ are factors (ranging from 0 to 1) that describe the influence of pH and temperature. These factors are typically derived experimentally as functions of optimal pH (pH_{opt}) and the limit (pH_{min} and pH_{max}) where growth ceases [15]. Similar relationships apply for temperature. Although these factors are often ignored for simplicity—since conditions are usually maintained near optimum—they become important in large-scale production where homogenization is challenging. In such cases, Euler–Lagrange computational fluid dynamics (CFD) can be used to describe the trajectories of individual cells and subpopulations [16].

The mass balance equation is also useful for describing control actions such as feed and base/acid addition. For example, when the batch process operates under perfusion mode with fresh medium and 100% cell retention, Equation (7.1) can be extended to

$$\frac{dX}{dt} = \mu X - FX \tag{7.7}$$

where F is the perfusion volume per unit time regarding the total volume.

The O_2/CO_2 transfer and biomass transfer are also described by mass balance equations. The transfer rate is influenced by bioreactor configuration like size, impeller, and spargers' type and position, as well as the property of the liquid phase such as viscosity. The control parameters also affect the transfer coefficient, for example, stirring rate affects the transfer coefficient of gas and heat in their balance equations, and the relation has been used to adaptively adjust dissolved oxygen.

Downstream modelling focuses on the kinetics of product recovery and impurity removal across several consecutive steps, most of which can be represented using mass balance equations. The performance of chromatography or filtration operations is influenced by material and geometric factors of the column or filter, buffer properties such as viscosity and ionic strength, and physicochemical variables during operation, including pH, temperature, and pressure. These factors are typically summarized into apparent parameters that must be determined experimentally. For example, when modelling the diafiltration of mRNA using tangential flow filtration, one can define the rejection factor of mRNA as follows:

$$R = 1 - \frac{C_p}{C_{FT}} \tag{7.8}$$

where R is the rejection factor, and C_p and C_{FT} are the mRNA in permeate and feed tank or before and after it went through the filter. Then, the concentration of mRNA in the feed tank can be expressed as follows:

$$C_{FT}(t) = C_{FT}(0)e^{D(R-1)t/V_{FT}} \tag{7.9}$$

where D is the buffer inlet rate, and V_{FT} is the volume of feed tank [17].

7.3.2 Data-Driven Models

Mechanistic models are valuable because they offer direct interpretation of a model by relating it to the biochemistry of a bioprocess. In many scenarios, a mechanistic model can be developed that serves as a reasonable approximation of a process under most conditions (but not all). However, in the context of PAT, many different sensors, probes, and spectra can be used to monitor several CPPs simultaneously. Here is where mechanistic models can reach certain limitations.

For instance, consider the mechanistic model proposed in Equation (7.3) for cell growth and death in the presence of a substrate and inhibitor. Intuitively, the equation makes sense. A high substrate concentration and low inhibitor concentration will maximize the specific growth rate of cells. The equation also includes limitations to growth and death, such that these rates cannot approach infinity. However, it is still, overall, an empirical formula since it is very difficult to break down the complete biochemistry and interaction between a given substrate, inhibitor, and cell type. This is due to the presence of countless other compounds found both inside and outside of the cells. In addition, there are other CPPs to consider such as temperature, pH, DO, and mixing speed which all greatly influence the cell growth rate.

With the advent of Industry 4.0, bioprocess modelling and PAT must adapt to the world of big data. Data-driven models do exactly that by learning the way that all the different measured process variables in a bioreactor can influence another parameter (e.g., cell growth rate). Using a data-driven approach, one no longer needs to hypothesize what the structure of a function will be to model cell growth rate based on substrate and inhibitor concentration, cell density, temperature (T), pH, DO, mixing speed (M), and any other measurable parameters. Such a model (f) would be abstract and represented initially as follows:

$$\frac{dX}{dt} = f\left(X, S, I, T, pH, DO, M\right) \tag{7.10}$$

The function f is a regression model that can take many forms, the most common of which were discussed in Section 5.3 of Chapter 5.

Although the convenience and the flexibility of data-driven models are clear, there also exist several disadvantages relative to mechanistic models. Since the model is not grounded in biological theory, the extrapolation of data from a data-driven machine learning model carries a high risk of error. For instance, an artificial neural network that was trained on data in a pH of 6.5–8.0 and a temperature of 32°C–38°C will typically have unreliable performance at a pH of 6 and temperature of 30°C since its neurons were not tuned to predict data in this region. Conversely, consider Equation (7.3) for cell growth with inhibitors. Although extrapolation always carries with it an increase in prediction error, safeguards exist in the form of μ_{max} and kd_{max} which keep the growth and death rate grounded within a reasonable range even when working outside the training data used to solve for these coefficients. Another limitation of data-driven models is their limited interpretability. The ease with which various machine learning models can be interpreted varies greatly. However, even when an interpretation is possible, it is not always straightforward to tie it back to biological theory.

7.3.3 Hybrid Models

We have discussed the advantages and disadvantages of both mechanistic models and data-driven models. Hybrid models are an attempt to capture the best of both worlds. As the name suggests, hybrid models combine mechanistic models and data-driven models. Another way of thinking about it is that a mechanistic model can be compartmentalized into data-driven components.

Let us consider once again the example of Equation (7.3). If we wanted to convert this mechanistic model into a hybrid model, we could (for example) do the following:

$$\frac{dX}{dt} = \left(\mu_{max} f(S,I) - kd_{max}\, g(S,I)\right) X \tag{7.11}$$

where f and g are data-driven components to a mechanistic model. However, to truly make this a hybrid model, f and g must experience the same limitations as their mechanistic counterparts. In this case, the mechanistic components that f and g replaced were restricted within the range of 0 and 1. One method to achieve this would be to use an artificial neural network that uses a sigmoid activation function in the last layer. This would allow the simultaneous prediction of f and g as a function of S and I (and any other variables of interest).

While the interpretability of this hybrid model is not as straightforward as that of the mechanistic equivalent in Equation (7.3), it can reveal patterns that could be used to reverse engineer a better mechanistic model. For instance, plotting a heat map of f and g as a function of S and I can reveal a pattern that may look similar to a more easily interpretable mathematical function. If a biological theory can be used to support the use of this mathematical function, then a new mechanistic model can be proposed as a result.

Hybrid models are a versatile, robust, and scalable type of model, which can be used in a variety of PAT applications. By breaking down a large mechanistic model into smaller data-driven components, one can refine one's modelling approach and gain new insights into the process that may offer enhanced interpretability.

7.4 PROCESS MODELLING AND DIGITAL TWINS

Biomanufacturing is increasingly transitioning towards autonomous and intelligent operation, driven by advances in sensing, computation, and model-based decision-making. Bioprocess digital twins (DT) are proposed as physical bioprocesses and their virtual replicas with continuous mutual data exchange. The DT aims to monitor, predict, optimize, and control the bioprocess—in real time and across development stages [18]. The physical component of DT includes the biomanufacturing facilities with sensors and controller as well as biological and physicochemical variables to be monitored. The virtual or digital component of DT is more or less the process models. Depending on the level of data exchange, Kritzinger et al. proposed a classification of the integration of DT into three levels: digital model (DM), digital shadow (DS), and DT [19]. A DM represents the system in isolation, without automated information exchange with the physical process. Such off-line models are primarily used to enhance process understanding and to perform simulation studies under hypothetical operating conditions. A DS extends the DM by incorporating one-way data flow from the physical process to the digital representation through sensors and PATs. The availability of real-time process data enables continuous state estimation, monitoring, and prediction. A DT further introduces bidirectional data exchange, allowing model outputs to inform control actions applied to the physical process. This capability supports autonomous operation through model-predictive control. This hierarchy implies that the DM forms the core of a digital twin, as the functional scope and capabilities of a DT are fundamentally determined by the scope and accuracy of the underlying model. For DT applications, the modelling depth is therefore limited by the CQAs and CPPs that can be measured or inferred through PAT in real time. As such, PAT is the fundamental layer of DT, which not only supplies model inputs but also defines the practical boundaries of DT.

REFERENCES

1. Penfold, W.J., and D. Norris, *The relation of concentration of food supply to the generation-time of bacteria.*. J Hyg, 1912. **12**(4): p. 527–531.
2. Noll, P., and M. Henkel, *History and evolution of modeling in biotechnology: Modeling & simulation, application and hardware performance.* Comput Struct Biotechnol J, 2020. **18**: p. 3309–3323.
3. FDA, Guidance for Industry: PAT—A Framework for Innovative Pharmaceutical Development, Manufacturing, and Quality Assurance. 2004.
4. Monod, J., *The growth of bacterial cultures.* Annu Rev Microbiol, 1949. **3**: p. 371–394.
5. Blackman, F.F., *Optima and limiting factors.* Annals of Botany, 1905. **os-19**(2): p. 281–296.
6. Teissier G., *Les lois quantitatives de la croissance.* Ann Physiol Physiochim Biol 1936(12): p. 527–573.
7. Moser, H., The Dynamics of Bacterial Populations Maintained in the Chemostat. 1958, Washington: Carnegie Institution of Washington.
8. Silva, C.A.T., A.A. Kamen, and O. Henry, *Intensified influenza virus production in suspension HEK293SF cell cultures operated in fed-batch or perfusion with continuous harvest.* Vaccines (Basel), 2023. **11**(12).
9. Sidoli, F.R., A. Mantalaris, and S.P. Asprey, *Modelling of mammalian cells and cell culture processes.* Cytotechnology, 2004. **44**(1–2): p. 27–46.
10. Henry, O., M. Perrier, and A. Kamen, *Metabolic flux analysis of HEK-293 cells in perfusion cultures for the production of adenoviral vectors.* Metab Eng, 2005. **7**(5–6): p. 467–476.
11. Okamura, K., et al., *Hybrid modeling of CHO cell cultivation in monoclonal antibody production with an impurity generation module.* Ind Eng Chem Res, 2022. **61**(40): p. 14898–14909.
12. Yang, O., and M. Ierapetritou, *mAb production modeling and design space evaluation including glycosylation process.* Processes, 2021. **9**: p. 324. DOI: 10.3390/pr9020324.
13. Hefzi, H., et al., *A consensus genome-scale reconstruction of Chinese hamster ovary cell metabolism.* Cell Syst, 2016. **3**(5): p. 434–443.e8.

14. Canova, C.T., P.K. Inguva, and R.D. Braatz, *Mechanistic modeling of viral particle production*. Biotechnol Bioeng, 2023. **120**(3): p. 629–641.
15. Rosso, L., et al., *Convenient model to describe the combined effects of temperature and pH on microbial growth*. Appl Environ Microbiol, 1995. **61**(2): p. 610–616.
16. Haringa, C., H.J. Noorman, and R.F. Mudde, *Lagrangian modeling of hydrodynamic–kinetic interactions in (bio)chemical reactors: Practical implementation and setup guidelines*. Chem Eng Sci, 2017. **157**: p. 159–168.
17. Nourafkan, E., et al., *An experimental and modeling approach to study tangential flow filtration performance for mRNA drug substance purification*. Biotechnol J, 2024. **19**(11): p. e202400473.
18. Portela, R.M.C., et al., When Is an in Silico Representation a Digital Twin? A Biopharmaceutical Industry Approach to the Digital Twin Concept Advances in Biochemical Engineering/Biotechnology, 2020.
19. Kritzinger, W., et al., *Digital twin in manufacturing: A categorical literature review and classification*. IFAC-PapersOnLine, 2018. **51**(11): p. 1016–1022.

8 Process Data Management

Javid Jannat and Mathieu Bourgey

8.1 GOOD RECORD MANAGEMENT PRINCIPLES

When gathering and storing data, it is essential to remember that the data is intended for use by people, machines, or software. It is important that data is managed and maintained accordingly. A fundamental principle is to keep different types of documents separate. For example, raw data, descriptions, procedures, and experimental notes should be kept in separate files while maintaining clear links between related items. This ensures the traceability of any given data file to the experiment or original source.

The most common and accessible way to store data is in tabular form, such as CSV files or SQL databases. When creating tables, a clean and consistent structure is important. Columns represent individual fields (e.g., date, value, unit), with field names included as column headers, typically in the first row. Each subsequent row corresponds to a single data record. All fields are populated for every row; if a valid value is unavailable, a placeholder such as "N/A" or "null" is used instead. When recording data in Excel or other spreadsheet software, each worksheet contains a single table positioned at the top-left corner, with no empty rows or columns above or to the left. This layout reduces confusion and helps ensure compatibility when converting the data to other formats or importing it into different software systems. Any calculations, annotations, or non-data notes are stored in separate sheets or files that can be clearly linked to the original data table.

Records are accompanied by metadata, which provides information and context for the data and facilitates management and traceability. Recorded data represent only a snapshot of a process at a given time and cannot be fully understood, analyzed, or trusted without the associated metadata. Key metadata elements include:

- ***Date and/or time*:** When the recording was made and when it was added to the database.
- ***Status*:** A brief description of the record's state (e.g., waiting, in progress, incomplete, in review, active, redacted, archived).
- ***Origin*:** The source of the record and relevant contacts (e.g., author, department, location, activity).
- ***ID*:** A unique alphanumeric code assigned to each record, enabling traceability and localization within the database. Data management operations are conducted through this ID, which is used for auditing, modifying records, and linking records together.

 DOI: 10.1201/9781032662961-8

For data management and identification purposes, the definition of a "record" and the way it is documented vary depending on the application. For example, when a machine measures a single parameter for a sample, that measurement can be treated as one record and assigned an ID. When multiple measurements are taken simultaneously, ambiguity may arise. If a machine measures pH and temperature, these measurements can be treated as separate data points with unique IDs. In contrast, an XYZ coordinate is more practically handled as a single data point and recorded as a vector. For more complex data, such as spectral analyses, a defined procedure for recording and handling the data is often required. For instance, when recording an IR spectrum, the raw data may be stored in a separate file and format (e.g., JCAMP-DX), while the file location and analytical conclusion (e.g., 99% toluene) are recorded in a table. These elements are linked using the record ID. This approach condenses complex data into a simplified tabular form, although it introduces a hierarchical structure. Establishing a consistent system for recording and storing data, along with documentation describing how the system operates, is critical. A consistent and standardized data structure supports clarity and usability; difficulty in clearly describing the structure often indicates that the data is disorganized and requires restructuring.

For data management purposes, GxP data (e.g., Good Manufacturing Practice [GMP], Good Clinical Practice [GCP], Good Laboratory Practice [GLP]) are commonly stored separately from non-GxP data. GxP standards impose strict regulations and procedures on how data and documents are managed, which require greater diligence, effort, and ongoing maintenance. These requirements can affect project timelines and costs and are generally unnecessary for non-GxP activities that are not subject to audits. In addition, mixing non-GxP and GxP data can create confusion and increase the risk of errors. Maintaining a clear separation between these data types supports compliance and reduces operational risk.

8.2 DATA LIFECYCLE MANAGEMENT

Depending on the field and use case, the data lifecycle can take different meanings and include different stages. Here, the seven most important stages of data lifecycle management [1, 2] are discussed (Figure 8.1).

8.2.1 Stage 1: Data Design

Before data collection begins, a plan is established to define the purpose, governance, and management of the data throughout its lifecycle, including the objectives of the bioreactor process. The plan documents critical process parameters (CPPs), critical quality attributes (CQAs), sensor sampling frequency, and acceptable operating ranges. It also identifies roles and responsibilities for recording, reviewing, and storing data, as well as the locations where the data is stored.

8.2.2 Stage 2: Data Creation

Data generation and collection can involve performing experiments, compiling existing data from external databases, or conducting surveys with patients or customers.

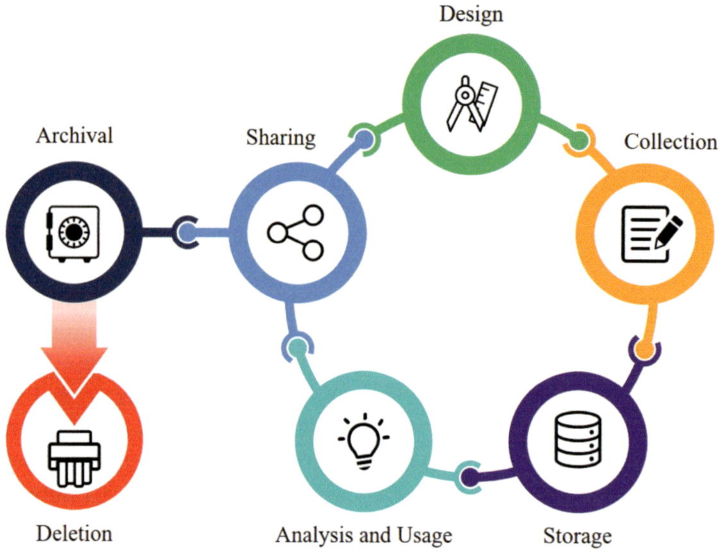

FIGURE 8.1 Flow of data throughout its lifecycle. First, a plan is established in the design phase. Data is then collected, stored, and prepared for analysis. Once data is analyzed and used to generate value, the results can be shared, which often drives new experimentation. Data that is deemed unfit for continued use is no longer shared and becomes archived. Archived data remains accessible until it is permanently deleted.

During this stage, data is collected in accordance with defined quality standards while meeting the required data volume. Failure to meet these standards can compromise data integrity and limit the ability to extract value, and in some cases may introduce risk to users. Conversely, exceeding quality standards or data quotas can lead to unnecessary costs and delays in development. Data capture is contemporaneous, with the time of measurement aligned with the time of recording. The acquisition process also preserves contextual information, including the sensor used, calibration status, batch, and operator, enabling traceability and reliable interpretation of the data.

8.2.3 Stage 3: Data Storage

At this stage, data is stored securely and prepared for use. The storage approach supports the accessibility needs of analysts and clients, as well as any safety or security requirements associated with the data. Decisions about data storage take into account data volume, security requirements, searchability, and accessibility for the intended user base. Online cloud services are commonly used because they support large data volumes, serve distributed users, and offer built-in encryption and access controls. For datasets with higher security requirements, locally accessible physical systems may be used instead. In addition, data may be preprocessed into specific formats to meet client or application requirements.

8.2.4 Stage 4: Data Analysis and Usage

This stage encompasses raw data cleaning, processing, analysis, and insight generation. Activities include data validation to ensure accuracy, trend analysis to identify patterns, and process optimization to improve efficiency and yield. This stage supports continuous improvement and process control, and it also contributes directly to data or product delivery to clients. Delivery commonly involves removing unnecessary information from the collected dataset and presenting the remaining data in a clear, legible format. When multiple clients are involved, differing requirements for content, formatting, and preprocessing are common. In such cases, data is typically formatted to be broadly legible and accessible to all parties, with any additional or specialized processing performed by the client as needed.

8.2.5 Stage 5: Data Sharing

In this stage, data is distributed to clients, and access controls are managed to ensure that authorized parties can retrieve the appropriate information. When datasets contain sensitive information, safeguards are applied to prevent unauthorized access. Sensitive portions may be redacted for certain recipients, or access may be limited to trusted parties through permission controls. Communication with clients confirms that data access is functioning as intended and that the data has not been compromised or corrupted during transfer. Data integrity can be verified using mechanisms such as checksum verification. System or user errors may occasionally restrict access, and these issues are addressed to ensure proper distribution. Ongoing maintenance of the data and its storage systems includes resolving access, integrity, and usability concerns as they arise.

8.2.6 Stage 6: Data Archival

In some cases, a dataset may be deemed no longer suitable for active use. Even then, retaining a copy of the data is important, whether due to ongoing projects that rely on legacy data, legal requirements, or audit purposes. Proper archival storage supports future audits and research. Long-term usability also depends on durable file formats and migration strategies. Digital records stored on hard drives or linear tape are commonly used, though the critical consideration is that data remain accessible with their context intact whenever needed. Data retention policies define how long data must be kept, the acceptable formats for storage, and the conditions under which data may be archived or destroyed.

8.2.7 Stage 7: Data Deletion

This stage represents the final step in the data lifecycle: data destruction. Many datasets have a maximum retention period, after which they must be securely destroyed, as is often the case for user or patient data. In regulated manufacturing environments, retention periods can extend five to ten years or longer. Alternatively, if a dataset no longer provides value and is not subject to regulatory retention requirements, it may be deleted. Datasets containing sensitive information require additional safeguards to prevent unauthorized access or reconstruction during destruction. In all cases, the destruction process is documented to meet regulatory and compliance requirements (Figure 8.1).

- ***Traceable*:** Data is tracked throughout their lifecycle, with a clear audit trail that records every addition, deletion, or modification while preserving the original entries. Traceability supports activities such as root cause analysis, enabling identification of which parameters changed, who made the changes, and the context in which they occurred.

Modification of records and error correction are permitted under GDP, but only under strict guidelines. For written entries, corrections must be made by drawing a single line through the error so that the original text remains legible. The correction is made next to the error and must include the date, the individual's signature, and the reason for the correction (often noted with an abbreviation such as EE for "entry error"). Organizations establish and document standard procedures for error correction, including a list of approved abbreviations for common mistakes. When space on a document is limited, a footnote may be used to provide clarification. For electronic records, an audit trail is maintained to ensure that the complete modification history of the record remains accessible.

GDP guidelines apply to more than just raw data, with their scope depending on the specific GxP standards relevant to the activity. In general, this includes all documents containing specifications, instructions, procedures, and records related to GxP activities. These procedures are managed within a documentation management system and are periodically reviewed to ensure they remain current, accurate, and clear, minimizing the risk of errors or misinterpretation.

8.4 SYSTEMS AND INFRASTRUCTURE IN PROCESS DATA MANAGEMENT

In modern bioprocessing environments, most process data is not recorded directly by individuals, but by computerized systems that link instruments to users. Although terminology and product names vary, generic system types are commonly used:

1. The SCADA (Supervisory Control and Data Acquisition) systems connect to controllers and sensors, acquire real-time data, and provide operators with dashboards and alarms. In bioprocessing, SCADA typically records time series such as temperature, pH, dissolved oxygen, agitation speed, gas flows, and other online measurements from bioreactors and utilities [11, 12].
2. MES (Manufacturing Execution Systems) orchestrate the execution of manufacturing steps, enforce recipes and workflows, and capture electronic batch records. MES often holds information on materials, instructions, setpoints, operator interventions, and the status of each batch or lot [13].
3. LIMS (Laboratory Information Management Systems) manage samples, analytical methods, and test results. LIMS link individual measurements to samples, batches, methods, instruments, and analysts, and are widely used in quality control and development laboratories [14, 15].

Data historians and data servers provide specialized storage for data from SCADA and controllers, optimized for long-term trending and retrieval. These systems do not exist in isolation. Together, they form part of a broader IT infrastructure that includes industrial networks (connecting controllers and SCADA), business networks (connecting MES, LIMS, and enterprise systems), and physical or cloud-based servers where databases and applications run.

From a data-management perspective, this infrastructure determines where data is born:

1. Where data is born (for example, a sensor via SCADA, an assay via LIMS, a step execution via MES).
2. How data travels between systems (direct interfaces, file exchanges, APIs).
3. Where authoritative records are stored, and which copies are considered secondary.

Throughout this chapter, we refer to SCADA, MES, LIMS, and related systems collectively as data providers and consumers. Understanding their roles and the underlying infrastructure is essential when designing identifiers, metadata, and integration strategies, and when assessing the feasibility and risks of any process data management solution

8.5 MANAGING PROVIDERS AND COMPLEX DATA

A data provider can be any person, system, or organization that contributes information to the process. This includes technicians and operators who perform experiments, calibrate equipment, or perform maintenance; medical professionals who collect patient and clinical trial data; vendors who supply reagents, consumables, equipment, or services; and external partners such as contract labs or manufacturing sites. In modern facilities, many providers are not humans but computerized systems: SCADA systems that acquire and supervise process data, MES that orchestrate manufacturing and capture electronic batch records, and LIMS that manage samples and analytical results. Process data can also come from public databases, regulatory or scientific literature, or from market research. For example, if a new product is being developed to meet specific consumer demands, the relevant consumer data must be documented. Similarly, if a process is being optimized based on findings from a scientific article, that source and its findings must also be recorded.

As soon as multiple providers are involved, even when data is "generated directly by the process," conflicts and inconsistencies can arise. For a complex process, different teams and equipment may be involved in each step, different types of measurements may be needed, and on any given day, a different technician may be recording the data. The same process may be performed at different facilities. Different departments (e.g., R&D and manufacturing) may record the same data differently; for example, the R&D lab may capture its data in spreadsheets, while the manufacturing department uses SCADA and MES. All these variables make consolidating and managing process data very difficult and complex.

This section addresses potential challenges that arise when integrating data from multiple, heterogeneous sources—including human and automated inputs, as well as internal and external providers—and strategies to mitigate these challenges to create a coherent dataset. Systematic use of unique identifiers, metadata, and critical attributes provides anchors across systems such as SCADA, MES, and LIMS, as well as across different locations, teams, and time periods.

8.5.1 Data Identity, Ownership, and Status

Each datapoint is assigned a unique identifier (ID), typically an alphanumeric code, to enable tracking, management, and linking across systems. When a datapoint originates from an experiment or process, the ID connects it to relevant documentation, such as batch records, protocols, or deviation reports. Each datapoint also includes clearly labeled ownership and status. The owner is the individual or group responsible for the datapoint and serves as the point of contact if issues arise. In many cases, this is the human provider; in automated systems, it may be a laboratory, department, or system instance (e.g., SCADA, MES, or LIMS) responsible for generating the data. The status provides a brief descriptor to flag datapoints requiring attention (e.g., draft, under review, approved, deprecated). Timeliness is a critical metadata element. Any recording, addition, modification, or deletion of a datapoint includes a timestamp, the identity of the person or system making the change, and the reason for the action. In validated systems such as MES or LIMS, this is maintained through audit trails. Depending on the context, datapoints may also include an expiration date, after which deletion or archiving is permitted.

8.5.2 Fragmentation and Conflicts

Data is often unintelligible between different organizations and teams, leading to consequences at operational [16], business [17], and patient care [18] levels. Each party often has unique goals, methodologies, and applications; they therefore record and store data in a manner that is optimized for their use case. For example, R&D and manufacturing departments may have different instruments and protocols for measuring pH because they handle different volumes, work at different scales, or use different SCADA/MES configurations. One site may rely heavily on LIMS and structured sample IDs, while another still uses semi-structured spreadsheets. This leads to fragmented data, which makes the dataset hard to consolidate and harmonize. The most important step is to enforce the data requirements and quality standards of the project or program. This will often require collaboration between the managers and the providers to transform the data into a form that is acceptable. To that purpose, it would require defining canonical units and naming conventions, mapping SCADA tags to MES parameters, or standardizing how LIMS test codes are represented. Even when under the same form, data from different sources may contradict one another. As such, metadata about the information source should be recorded. This includes information about the physical location where the data was recorded, the team that recorded it, as well as the protocol and the instruments used, and, where relevant, the originating system (SCADA node, MES recipe, LIMS project). This data should be present in the documentation linked to the ID, but it can also be incorporated into the table where appropriate.

8.5.3 Volatility, Hierarchical Structure, and Confidence

Certain subsets of data can pose unique challenges. Some data may be subject to frequent modifications; this could be caused by the very nature of the data [19] (for example, derived model parameters) or because it is owned and used by many different parties within the organization [20]. Configuration data in SCADA, master data in MES, or method definitions in LIMS are good examples of such volatile data. Volatility can impact data quality and confidence. The previously discussed recommendations still apply to this volatile data. Though it is important to assess the associated risks, if a subset of data is too volatile to be used in a project or if it is deemed too difficult to manage, then it should be isolated and moved to a dataset designed to store volatile data or handled through stricter governance. Generally, a dataset will only contain a single copy of each datapoint. With volatile data, however, it is mandatory to make a new copy of the data every time it is modified, to keep the original and previous versions. This means that the set will contain multiple entries with the same ID. To help with clarity, only one entry should be labeled as active, old entries should be labeled as deprecated, and future entries should be labeled as awaiting validation and have a comment explaining the modification. This way, if data used in a report becomes modified, that data and its version history are readily available.

Some experiments or protocols may produce data that are interrelated or dependent on one another [21]. For instance, a substance is confirmed to be ethanol via spectroscopy; the veracity of that claim depends on the spectroscopic data, which in turn depends on the calibration data of the instrument. In bioprocessing, a product titer may depend on upstream bioreactor conditions recorded in SCADA, which depend on controller settings defined in MES, which depend on raw material information stored in LIMS. This is an example of hierarchical data. For very simple hierarchies, the dependencies can be linked in the experiment report and linked to the ID. More complex hierarchies need to be planned for [22], and ideally, every entry in the dataset should have the same hierarchical structure. That structure, the relationship between each node (for example, connecting batch to unit operation and unit operation to analytical results), as well as where to find each node (which system, which table), should be clearly documented. As the complexity and variety of hierarchies increase, the dataset becomes increasingly difficult to understand. If two providers provide the same type of data with very different structures, then their data should be kept in separate sets or harmonized through an intermediate layer.

Hierarchical data tends to be more volatile: if one datapoint is modified, all data related to it needs to be revised [23]. Data reliability and confidence suffer from this problem. Monitoring data confidence is therefore important when data points are hierarchical and affect one another, which is often the case in process data. Confidence generally measures the overall accuracy and precision of a measurement. This can be expressed as variance, confidence intervals, p-values, or any metrics that represent the level of confidence in the measurement. The metric depends on the application: for yield data that is used to calculate an average, this could be instrument error and variance; for spectroscopic data used to measure concentration, this could be the confidence interval of the calibration model.

8.5.4 Limitations in Data Handling

Managing complex data from multiple sources and providers while maintaining quality and usability can be challenging. At the start of a project, data engineers collaborate with clients, assess data providers, and determine the types of data to be collected and managed. Based on this analysis, architecture is developed to support these requirements. In modern environments, this often involves explicitly defining how systems such as SCADA, MES, LIMS, and others will contribute to and consume shared data structures. The architecture is ideally designed to accommodate future improvements and the recording of additional information, such as comments, descriptions, or new attributes. For example, the integration of a new process analytical technology (PAT) analyzer, the addition of a LIMS module, or the onboarding of an external contract laboratory may be required. Architectural limitations are explicitly identified. If evolving data requirements exceed current capabilities, data may be split into separate sets (e.g., separating legacy data from new structured data), or the database and integration layer may be redesigned to meet new demands. Making these trade-offs explicit allows organizations to develop their data landscape in a controlled and scalable manner, avoiding the accumulation of opaque or incompatible data fragments over time.

8.6 CLIENT MANAGEMENT AND DATA DELIVERY

While much of this discussion has focused on the upstream aspects of data management—such as data generation, record organization, provider and system coordination, and storage—the downstream aspects are equally important. Downstream management addresses how data is delivered to clients and how client requirements influence the design of data systems. A data client is any party that uses data to make decisions or perform tasks. In a bioprocess environment, clients include teams or personnel involved in process development and engineering, manufacturing operations, quality control and assurance, regulatory verification, pharmacovigilance, and business or management operations. External parties, such as contract manufacturing organizations, contract research organizations, or regulatory authorities, are also considered clients. Each client may have specific expectations regarding the granularity, timeliness, format, and contextual information of the data they receive. Understanding these requirements is essential for designing data systems that deliver value while maintaining consistency, traceability, and usability.

It is important to consider data availability as a service from a management perspective [24]. Data managers collaborate with stakeholders to define key data management characteristics for each major client class or use case:

- The **scope** of data to be delivered (which variables, which time periods, which batches).
- The **required level of detail** (raw time series versus batch summary statistics).
- The **acceptable latency** (near real-time, daily, monthly, on demand).
- The **format and interface** (files, dashboards, APIs, reports).
- The **constraints** related to data protection and confidentiality.

A clear definition of these parameters helps to reduce error and increase efficiency, avoiding unplanned manual data extractions. It also helps to preserve data security by controlling unplanned access to raw systems that may compromise data integrity and confidentiality.

Data delivery is a regulatory concern that significantly impacts the GxP capability [25]. Many key deliverables (batch release packages, annual product quality reviews, and regulatory submissions) depend on data that must be complete, accurate, traceable, and consistent with the validated state of the process. Thus, to support the regulatory procedure, any automated extraction or transformation that feeds these deliverables should therefore be documented, controlled, and traceable. For example, data managers maintain visibility into what data is extracted, how they are transformed, and where they are stored. Any modifications are assessed for impact and approved according to predefined change management protocols. It must always be possible to reconstruct which data, including specific versions, as well as the tools and procedures used, contributed to any report or decision. These requirements align with those applied to laboratory methods or manufacturing processes. The same approach is applied when data from process analyzers and sensors is used in real-time control or real-time release.

As organizations collect more data and use more advanced analytical processes, there is a developing interest in self-service access solutions [26, 27]. Thus, scientists, engineers, and analysts want to explore data directly using their specific and preferred statistical tools, programming languages, or notebooks, without relying on extracts from data managers. This new trend may accelerate process understanding and result delivery, but it comes with new specific challenges to address. Initially, data managers restrict access to curated or governed datasets to preserve the integrity of the raw data. Data clients are provided with clear information about the origin, limitations, and context of the data they access, ensuring that data lineage and quality indicators remain visible. Clients also receive training on self-service access processes and tools, as well as guidance on the content and metadata available to them.

This approach reflects practices commonly seen in academic or early-stage development environments, where flexibility and learning are prioritized. As development progresses toward formal validation and commercial manufacturing, data delivery typically becomes more standardized, automated, and closely integrated with the overall quality system.

By maintaining data integrity and complying with legal restrictions, good client management and data delivery practices aim to ensure that the right individuals receive the correct data in the appropriate format at the right time.

8.7 DATA QUALITY

Data quality is an abstract and multifaceted concept. Depending on the application and the relevant regulatory framework, it is typically broken down into a set of more specific and measurable dimensions. Some of the more common dimensions [28–30] are listed below:

- ***Findability***: The process of making data discoverable beyond the people who produced it. Therefore, data and metadata have unique identifiers and are indexed to be located later. This requires the use of structured folders with stable naming and a catalogue. In the industry, this relates to using batch IDs and searchable historians.

- ***Accessibility*:** Data can be retrieved using standardized methods, with authentication when appropriate. Accessible does not mean open to everyone; it means the access process is clear and reproducible. Accessibility removes bottlenecks to validation and innovation.
- ***Interoperability*:** Data should be usable across tools and systems relying on common formats, shared vocabularies, and clear units. In bioproduction, upstream, downstream, analytics, and QC systems must talk to each other. Interoperability is essential for AI because models need datasets merged without heroic data cleaning.
- ***Reusability*:** Data must have a clear license, their provenance must be known (how, when, and why data was generated), and they must comply with standards. In the industry, it supports trending, continuous improvement, and platform learning. Reusable data gives enough context that someone else can interpret and trust it without guessing.
- ***Timeliness*:** When was the record requested, when was the measurement made, when was the record added to the database, and if data is recorded at regular intervals, how often? One key application of timeliness is in quality control: detecting quality issues early means reducing damages and costs caused by a failed batch.
- ***Consistency*:** How similar is the data across operators, instruments, and records? Ideally, every entry in the database should follow the standard structure. Measurements of the same type should use the same units; when measuring the same parameter, different operators and instruments should obtain similar results; a consistent naming convention should be followed, standard operating procedure (SOP) and standard calibration techniques should be used and shared between the different parties of the same facility.
- ***Completeness*:** How much information is missing from the records that should be present? Ideally, all samples should have measurements for all requested parameters. Sometimes, a measurement does not make sense (e.g., pH of a solid); this is acceptable as long as it is noted that the measurement is non-applicable. Completeness refers to items that are explicitly incomplete or left empty.
- ***Accuracy*:** How well does the data reflect the real-world phenomena it is intended to represent? In scientific and technical contexts, accuracy depends on instrument performance, calibration, experimental design, and sample handling. Accurate data tends to minimize bias and random error. Accuracy relies on the use of clearly defined statistical methods, units, and uncertainty measurements.
- ***Reliability*:** How well is the data and its context preserved over time? Reliable data is well documented, stored in controlled systems, and accompanied by enough metadata to understand how they were produced and how they have changed. Reliability depends on the trustworthiness of the source, such as validated instruments, accredited labs, or controlled information systems, and it allows for diagnosing and resolving data quality issues.

Consequently, many of the concepts in data quality are closely related to GDP principles. In fact, each element of ALCOA++ can be viewed as its own quality dimension. It is important to note that ALCOA++ provides regulatory guidelines for data integrity and ensuring the data is acceptable in audits, inspections, or client sharing. In practice, data quality also needs to support reuse and value creation: how easily data can be found, accessed, combined with other datasets, or reused in new analyses. This reflects a broader perspective and aligns with the FAIR principles (making data *Findable, Accessible, Interoperable, and Reusable), a common concept in academia that is also being applied in industry* When choosing and defining quality dimensions, we need to ensure they are specific and relevant to the project and cover both regulatory requirements (ALCOA++) and client or internal needs for reuse (FAIR).

To assess data quality across various dimensions, it is useful to define and track specific quality attributes (QAs) [31]. A QA can be any characteristic of the data that describes its quality. Attributes typically fall into two categories: indicators and parameters [31]. Indicators are characteristics that can be directly calculated or obtained from the dataset (for example, the proportion of missing values, the date of the last measurement, or the identity of the instrument). Parameters, on the other hand, represent the characteristics that a client wants to know about the data (for example, "Is this data recent enough for my purpose?" or "Can I trust this measurement in a regulatory submission?"). In some cases, these parameters can be derived directly from the indicators, but more often they must be inferred from a combination of indicators or estimated using professional judgment.

8.7.1 Navigating Data Quality

For longer projects with extensive data requirements, establishing a data management strategy early is essential. Changing standards or requirements midway can create significant backlogs and rework. Stakeholders define the data quality requirements upfront, including key parameters and indicators, with deliberate overlap across dimensions to improve robustness and reliability [32, 33]. However, each dimension requires special considerations.

8.7.1.1 Timeliness

In a scientific context, how fast data can be made available depends heavily on the assay and the available technology [34]. A significant amount of scientific data tends to be very large and involves complex calculations. Wherever possible, data collection and processing should be computerized and automated.

Data also has a temporal context. Standards, procedures, and technologies evolve over time; as such, data collected ten years ago may not be as relevant as data collected this year.

8.7.1.2 Accessibility

Accessibility has two aspects: technical [35, 36] and cognitive [37]. On a technical level, clients can retrieve the data they need without complicated workarounds and receive notifications when new data is available. This typically involves searchable storage, clear links between datasets and documentation (such as protocols, calibration data, and reports), and well-defined interfaces or views. This allows auditing and verifying the data as well as observing the downstream effects should we need to modify or delete the data. Cognitively, the data must be comprehensible to users with diverse backgrounds.

Using standardized formats and reusable transformations (scripts, pipelines, or templates) customized for each client is more sustainable than manual adjustments.

8.7.1.3 Consistency and Completeness

Consistency and completeness are very similar from a management perspective [38–40]. Incomplete data means some information is missing; inconsistent data means what is present does not align. Inconsistencies can be fixed, but it becomes more difficult to work with as a database grows. Missing data cannot be corrected. However, they are often unavoidable but should be explicitly marked and, where possible, explained.

To reduce Consistency and Completeness impact, it is best to intercept them as early as possible; to train providers on naming and recording conventions; to configure systems such as LIMS, MES, e-forms to perform validation of entries at the point of capture; and to standardize the integration of third-party data before it enters core datasets.

8.7.1.4 Accuracy and Reliability

Data accuracy heavily depends on the data providers, methodology, and technology. From a data management standpoint, an attempt can be made to preserve the data as faithfully as possible and perform statistical analyses. It is also essential to document any protocol, instrument, or methodological limitations and incorporate these considerations into the statistics.

Still, data accuracy is a very complex phenomenon that has cascading effects throughout an organization. Database technology expert Jack E. Olson has written an excellent book on the subject [41]. When inaccuracies emerge, there are significant challenges in detection and diagnosis. It is therefore important to thoroughly integrate reliability into the database and follow good data management practices. Maintaining complete documentation of data, its source, and its version history is fundamental to tracking accuracy, auditing the data, and retroactively resolving issues [42, 43].

8.7.2 Data Profiling and Cleansing

As data begins to accumulate, active monitoring becomes essential to ensure its quality. Data profiling provides a structured way to examine datasets before they are used for analysis, modeling, or decision-making. Profiling tasks are generally separated into three categories [44, 45].

- *Single column profiling*: Examine each variable separately. This is used to analyze a variable's behavior throughout a dataset and is most useful for assessing consistency and completeness. Typical checks include the following:
  - *Profiling data type and domain*: What data type is stored in the variable, and what does it represent? For example, "11-Oct-2024" is a text-based data type, and it represents a date; "521.44$" is a numeric data type that represents monetary value. Ideally, all the data in a column contains the same data type.
  - *Data format*: How is the value represented? The dates "11-Oct-2024" and "11-10-2024" are different representations of the same

value. Ideally, all the values in a column should have the same representation.
    - *Value distribution and cardinality*: This depends on the data type. For numeric values, statistics about the distribution can be calculated. For text-based data types, the number of characters can be analyzed.
    - *Hapax detection*: How many values in a column are unique? Depending on the column, a hapax could represent a value that was incorrectly entered.
    - *Completeness*: What percentage of the entries in a column contain valid values? Depending on the application, a cleansing step can be used to fill in the missing values.
- *Multi-column profiling*: It looks at relationships between variables to reveal patterns invisible at the single-column level. This typically involves measuring correlations between columns, analyzing data using clustering algorithms, and plotting columns with respect to one another. Multi-column profiling is resource-intensive but is critical for understanding whether data can be pooled across providers and for detecting latent issues such as instrument drift.
- *Dependency profiling*: Dependency is an advanced subject, but generally, it deals with finding independent variables in a data set. In our case, this is the ID value. Using only the ID, we can find all the other information about an entry, such as a variable, which is often referred to as a key. Clear keys are essential for maintaining referential integrity when integrating SCADA, MES, LIMS, and other sources.

Once profiling is complete and issues have been identified, data cleansing can be initiated. It is important to note that cleansing cannot resolve all issues. Assays where data is missing or incorrectly recorded should be flagged as such and temporarily removed from the database until a new assay is performed. Typically, cleansing greatly improves consistency and accessibility and should be performed in four steps [46, 47]:

1. ***Define transformation rules***: A cleansing schema is defined in a database management software (e.g., transform all dates to dd-mm-yyyy format or transform all measurements from kg to g).
2. ***Test on a subset***: This step verifies that the schema performs the task as intended. It is advisable to ensure the schema does not corrupt correct data and can handle discrepancies without data loss.
3. ***Apply and validate***: Executing the transformation and creating a cleansed copy of the data. It is important to reprofile the cleansed data set and ensure quality metrics have improved and that there is no data loss before proceeding.
4. ***Replacing***: The original data should not be deleted. The best practice is to archive the data while maintaining a clear audit trail of how the data was manipulated and why. The replaced data can either be stored in the same table and labeled as "archived" or moved to a dedicated data set elsewhere.

Cleansing is particularly effective for removing duplicates, fixing spelling or formatting errors, and eliminating illegal or structurally invalid values. However, in scientific contexts, caution is required: cleansing items without a well-justified rationale can introduce bias and hurt accuracy. This is particularly important for tasks that involve outlier removal (Figure 8.2).

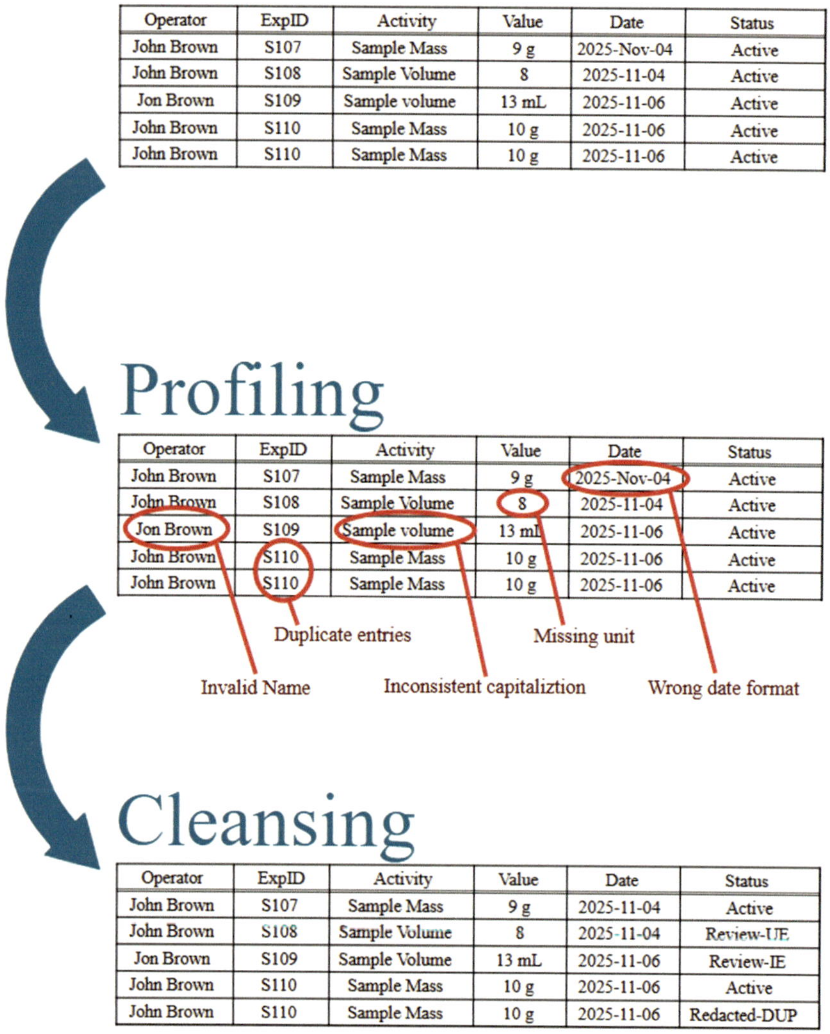

Operator	ExpID	Activity	Value	Date	Status
John Brown	S107	Sample Mass	9 g	2025-Nov-04	Active
John Brown	S108	Sample Volume	8	2025-11-04	Active
Jon Brown	S109	Sample volume	13 mL	2025-11-06	Active
John Brown	S110	Sample Mass	10 g	2025-11-06	Active
John Brown	S110	Sample Mass	10 g	2025-11-06	Active

Operator	ExpID	Activity	Value	Date	Status
John Brown	S107	Sample Mass	9 g	2025-Nov-04	Active
John Brown	S108	Sample Volume	8	2025-11-04	Active
Jon Brown	S109	Sample volume	13 mL	2025-11-06	Active
John Brown	S110	Sample Mass	10 g	2025-11-06	Active
John Brown	S110	Sample Mass	10 g	2025-11-06	Active

Operator	ExpID	Activity	Value	Date	Status
John Brown	S107	Sample Mass	9 g	2025-11-04	Active
John Brown	S108	Sample Volume	8	2025-11-04	Review-UE
Jon Brown	S109	Sample Volume	13 mL	2025-11-06	Review-IE
John Brown	S110	Sample Mass	10 g	2025-11-06	Active
John Brown	S110	Sample Mass	10 g	2025-11-06	Redacted-DUP

FIGURE 8.2 Example of a profiling and cleansing workflow. Profiling the data allows detection of several discrepancies in the data set: Sample 107 has an incorrect date format; 108 is missing the unit of measurement; 109 has a typo in the operator's name and inconsistent capitalization; and 110 is duplicated. In the cleansing step, the capitalization and date format are fixed without data loss. The duplicated entry is marked as redacted with the code DUP, indicating that the row is a duplicate. The last discrepancies require human verification. Sample 108 is flagged for review with code "UE", indicating a unit error. Sample 109 is flagged with code "IE", indicating an invalid entry.

8.8 TOWARD DATA LAKES FOR COMPLEX BIOPROCESS DATA

Most databases are structured as data warehouses where data is cleaned and structured before being stored in a central repository as large tables that are optimized for searching and navigation [48]. This is not ideal in scientific contexts where we often want the raw, unaltered files obtained from instruments and machines for analysis [49, 50]. For this reason, data lakes are preferred. A data lake is a centralized repository that stores large volumes of data in its native format until it is needed for processing and analysis. This includes data from SCADA and historians, MES, LIMS, and electronic laboratory notebook (ELN). Because the lake does not impose a rigid schema upfront, it is naturally more adaptable to bioprocesses that produce heterogeneous data and to R&D projects where both data and questions change rapidly. Bioprocess environments generate massive, continuous streams of data from many sources: upstream and downstream equipment, sensor-rich PAT tools, batch records, electronic lab notebooks, QC assays, and operator inputs. These sources differ widely in format, size, and frequency. A data lake is preferred in this context because it can store all data formats, including time series data, images, and lab logs. This eliminates overhead and allows raw data to be shared and used [49, 50].

Data lakes can also strengthen traceability and compliance, which are essential in GxP environments. Because data is stored in its raw form, teams preserve full lineage for every acquired dataset, including metadata, timestamps, and original source files. This supports auditability and troubleshooting. Combined with proper data governance, data lakes help ensure regulatory and GxP compliance while enabling flexible analytics [51, 52].

From a scientific and engineering perspective, data lakes facilitate deeper analytics and modeling. Data lakes provide the substrate for AI and digital twins [53–55]. AI and machine learning benefit from access to large, diverse, and high-granularity datasets. A lake can feed feature engineering, model training, and continuous model monitoring with minimal manual intervention. Instead of extracting narrow, project-specific subsets, teams can systematically reuse and recombine historical data for new models and use cases. Building a digital twin, which is a computational representation of the process used for simulation, prediction, or decision support, will require a structural model of the bioprocess and a continuous flow of real-world data to calibrate, update, and validate the model. A well-governed data lake makes it feasible to develop such a digital model synchronized with the actual process. This will ultimately support scenario analysis, what-if simulations, and AI-assisted optimization and control.

These benefits do come at a cost; however, due to the heterogeneous, hierarchical, and fast-changing nature of the data, data lakes are much harder to manage and monitor for quality issues. In a data warehouse, data is stored in tables and is severely limited in structure and format. This makes profiling and cleansing large quantities of data easy and straightforward. For this reason, QA data, CPPs, CQAs, and other sensitive data should be subject to strict standards with the goal of facilitating data management. Furthermore, data lakes can make collaboration between teams more difficult. As teams develop increasingly different data structures and data handling methods, it becomes more difficult to share data and communicate.

Overall, data lakes offer a powerful foundation for handling complex bioprocess data and for enabling advanced analytics, AI, and digital twins, but they require careful design, robust governance, and clear quality standards from a data management perspective.

REFERENCES

1. K. Rahul and R. K. Banyal, "Data life cycle management in big data analytics," *Procedia Comput. Sci.*, vol. 173, pp. 364–371, Jan. 2020, doi: 10.1016/j.procs.2020.06.042.
2. K. Miller, M. Miller, M. Moran, and B. Dai, "Data Management Life Cycle, Final Report." Accessed: Dec. 10, 2025. [Online]. Available: https://rosap.ntl.bts.gov
3. T. Harris, J. R. Larus, and R. Rajwar, *Transactional Memory*. California, USA: Morgan & Claypool Publishers, 2010.
4. WHO, "TRS 1033 – Annex 4: WHO Guideline on Data Integrity," 2023. Accessed: Dec. 12, 2025 [Online]. Available: "https://www.who.int/publications/m/item/annex-4-trs-1033" https://www.who.int/publications/m/item/annex-4-trs-1033
5. European Commission, "*EudraLex – Volume 4*," Public Health. Accessed: Apr. 22, 2026 [Online]. Available: https://health.ec.europa.eu/medicinal-products/eudralex/eudralex-volume-4_en
6. U.S. Food and Drug Administration, "*Current Good Manufacturing Practice in Manufacturing, Processing, Packing, or Holding of Drugs; General*," Code of Federal Regulations, Title 21, Part 210. Accessed: Dec. 12, 2025 [Online]. Available: "https://www.ecfr.gov/current/title-21/part-210" https://www.ecfr.gov/current/title-21/part-210
7. U.S. Food and Drug Administration, "*Current Good Manufacturing Practice for Positron Emission Tomography Drugs*," Code of Federal Regulations, Title 21, Part 212. Accessed: Dec. 12, 2025 [Online]. Available: "https://www.ecfr.gov/current/title-21/part-212" https://www.ecfr.gov/current/title-21/part-212
8. H. Canada, "Good Manufacturing Practices Guide for Drug Products (GUI-0001)." Accessed: Dec. 12, 2025. [Online]. Available: https://www.canada.ca/en/health-canada/services/drugs-health-products/compliance-enforcement/good-manufacturing-practices/guidance-documents/gmp-guidelines-0001/document.html
9. A. K. Rattan, "Data integrity: History, issues, and remediation of issues," *PDA J. Pharm. Sci. Technol.*, vol. 72, no. 2, pp. 105–116, Mar. 2018, doi: 10.5731/pdajpst.2017.007765.
10. H. Foldal, "Digitalization in the metrology quality infrastructure – perspectives from Novo Nordisk," *EPJ Web Conf.*, vol. 323, p. 01002, 2025, doi: 10.1051/epjconf/202532301002.
11. D. Upadhyay and S. Sampalli, "SCADA (Supervisory Control and Data Acquisition) systems: Vulnerability assessment and security recommendations," *Comput. Secur.*, vol. 89, p. 101666, Feb. 2020, doi: 10.1016/j.cose.2019.101666.
12. D. J. Gaushell and H. T. Darlington, "Supervisory control and data acquisition," *Proc. IEEE*, vol. 75, no. 12, pp. 1645–1658, Dec. 1987, doi: 10.1109/PROC.1987.13932.
13. B. Saenz de Ugarte, A. Artiba, and R. Pellerin, "Manufacturing execution system – a literature review," *Prod. Plan. Control*, vol. 20, no. 6, pp. 525–539, Sep. 2009, doi: 10.1080/09537280902938613.
14. K. Boyar, A. Pham, S. Swantek, G. Ward, and G. Herman, "Laboratory Information Management Systems (LIMS)," in *Cannabis Laboratory Fundamentals*, S. R. Opie, Ed., Cham: Springer International Publishing, 2021, pp. 131–151. doi: 10.1007/978-3-030-62716-4_7.
15. P. J. Prasad and G. L. Bodhe, "Trends in laboratory information management system," *Chemom. Intell. Lab. Syst.*, vol. 118, pp. 187–192, Aug. 2012, doi: 10.1016/j.chemolab.2012.07.001.

16. H. van der Aa, H. Leopold, F. Mannhardt, and H. A. Reijers, "On the Fragmentation of Process Information: Challenges, Solutions, and Outlook," in *Enterprise, Business-Process and Information Systems Modeling*, K. Gaaloul, R. Schmidt, S. Nurcan, S. Guerreiro, and Q. Ma, Eds., Cham: Springer International Publishing, 2015, pp. 3–18. doi: 10.1007/978-3-319-19237-6_1.
17. U. K. R. Gangula, "Streamlining data management in salesforce: From fragmentation to centralization," *J. Artif. Intell. Mach. Learn. Data Sci.*, vol. 3, no. 3, pp. 2783–2787, Jul. 2025, doi: 10.51219/JAIMLD/uday-kumar-reddy-gangula/584.
18. J. Puustjärvi and L. Puustjärvi, "Managing fragmented personal data: going beyond the limits of personal health records," presented at the *9th International Conference on Health Informatics*, Dec. 2025, pp. 145–150. Accessed: Dec. 12, 2025. [Online]. Available: https://www.scitepress.org/Link.aspx?doi=10.5220/0005626101450150
19. A. Gal and J. Eckstein, "Managing Periodically Updated Data in Relational Databases: A Stochastic Modeling Approach," Jun. 11, 2001, *arXiv*: arXiv:cs/0007044. doi: 10.48550/arXiv.cs/0007044.
20. H. Roitman, A. Gal, and L. Raschid, "Satisfying complex data needs using pull-based online monitoring of volatile data sources," in *2008 IEEE 24th International Conference on Data Engineering*, Apr. 2008, pp. 1465–1467. doi: 10.1109/ICDE.2008.4497591.
21. N. G. Saleri, R. M. Toronyi, and D. E. Snyder, "Data and data hierarchy," *J. Pet. Technol.*, vol. 44, no. 12, pp. 1286–1293, Dec. 1992, doi: 10.2118/21369-PA.
22. D. C. Tsichritzis and F. H. Lochovsky, "Hierarchical data-base management: a survey," *ACM Comput. Surv.*, vol. 8, no. 1, pp. 105–123, Mar. 1976, doi: 10.1145/356662.356667.
23. T. I. Korotkova and A. A. Mokhov, "Hierarchical database management model with uncertain parameters," *IOP Conf. Ser. Mater. Sci. Eng.*, vol. 927, no. 1, p. 012046, Sep. 2020, doi: 10.1088/1757-899X/927/1/012046.
24. L. Tedersoo *et al.*, "Data sharing practices and data availability upon request differ across scientific disciplines," *Sci. Data*, vol. 8, no. 1, p. 192, Jul. 2021, doi: 10.1038/s41597-021-00981-0.
25. S. C. Hock, V. Tay, V. Sachdeva, and C. L. Wah, "Pharmaceutical data integrity: issues, challenges and proposed solutions for manufacturers and inspectors.," *Generics Biosimilars Initiat. J.*, vol. 9, no. 4, pp. 171–183, Dec. 2020.
26. R. Ooms and M. Spruit, "Self-service data science in healthcare with automated machine learning," *Appl. Sci.*, vol. 10, no. 9, p. 2992, Jan. 2020, doi: 10.3390/app10092992.
27. P. Alpar and M. Schulz, "Self-service business intelligence," *Bus. Inf. Syst. Eng.*, vol. 58, no. 2, pp. 151–155, Apr. 2016, doi: 10.1007/s12599-016-0424-6.
28. Government of Canada, "*Guidance on Data Quality*," Canada.ca. Accessed: Dec. 12, 2025 [Online]. Available: https://www.canada.ca/en/government/system/digital-government/digital-government-innovations/information-management/guidance-data-quality.html
29. F. Sidi, P. H. Shariat Panahy, L. S. Affendey, M. A. Jabar, H. Ibrahim, and A. Mustapha, "Data quality: a survey of data quality dimensions," in *2012 International Conference on Information Retrieval & Knowledge Management*, Mar. 2012, pp. 300–304. doi: 10.1109/InfRKM.2012.6204995.
30. M. D. Wilkinson *et al.*, "The FAIR guiding principles for scientific data management and stewardship," *Sci. Data*, vol. 3, no. 1, p. 160018, Mar. 2016, doi: 10.1038/sdata.2016.18.
31. R. Y. Wang, M. P. Reddy, and H. B. Kon, "Toward quality data: an attribute-based approach," *Decis. Support Syst.*, vol. 13, no. 3–4, pp. 349–372, Mar. 1995, doi: 10.1016/0167 9236(93)E0050-N.
32. R. Y. Wang, H. B. Kon, and S. E. Madnick, "Data quality requirements analysis and modeling," in *Proceedings of IEEE 9th International Conference on Data Engineering*, Apr. 1993, pp. 670–677. doi: 10.1109/ICDE.1993.344012.

33. F. R. S. Paim and J. F. B. de Castro, "DWARF: an approach for requirements definition and management of data warehouse systems," in *Proceedings. 11th IEEE International Requirements Engineering Conference, 2003.*, Sep. 2003, pp. 75–84. doi: 10.1109/ICRE.2003.1232739.
34. J. H. Howanitz and P. J. Howanitz, "Laboratory results: timeliness as a quality attribute and strategy," *Am. J. Clin. Pathol.*, vol. 116, no. 3, pp. 311–315, Sep. 2001, doi: 10.1309/H0DY-6VTW-NB36-U3L6.
35. P. R. Baldwin *et al.*, "Big data in cryoEM: automated collection, processing and accessibility of EM data," *Curr. Opin. Microbiol.*, vol. 43, pp. 1–8, Jun. 2018, doi: 10.1016/j.mib.2017.10.005.
36. M. Velasco, J. M. Fuertes, and P. Marti, "Process data abstraction/accessibility via internet," in *EFTA 2003. 2003 IEEE Conference on Emerging Technologies and Factory Automation. Proceedings (Cat. No.03TH8696)*, Sep. 2003, pp. 357–363, vol. 1. doi: 10.1109/ETFA.2003.1247728.
37. S. Horton and D. Sloan, "Accessibility in Practice: A Process-Driven Approach to Accessibility," in *Inclusive Designing*, P. M. Langdon, J. Lazar, A. Heylighen, and H. Dong, Eds., Cham: Springer International Publishing, 2014, pp. 105–115. doi: 10.1007/978-3-319-05095-9_10.
38. Y. Timmerman, R. Nasfi, G. De Tré, F. Pattyn, and A. Bronselaer, "Cost-based analysis of the impact of data completeness and representational consistency," *Decis. Support Syst.*, vol. 175, p. 114044, Dec. 2023, doi: 10.1016/j.dss.2023.114044.
39. M. B. Aron *et al.*, "Towards improving district health information system data consistency, report completeness and timeliness in Neno district, Malawi," *BMC Med. Inform. Decis. Mak.*, vol. 24, no. 1, p. 376, Dec. 2024, doi: 10.1186/s12911-024-02802-x.
40. D. P. Ballou and H. L. Pazer, "Modeling completeness versus consistency tradeoffs in information decision contexts," *IEEE Trans. Knowl. Data Eng.*, vol. 15, no. 1, pp. 240–243, Jan. 2003, doi: 10.1109/TKDE.2003.1161595.
41. J. E. Olson, *Data Quality: The Accuracy Dimension*. California: Morgan Kaufmann Publisher, 2003.
42. N. Agmon and N. Ahituv, "Assessing data reliability in an information system," *J. Manag. Inf. Syst.*, vol. 4, no. 2, pp. 34–44, Sep. 1987, doi: 10.1080/07421222.1987.11517792.
43. R. Krishnan, J. Peters, R. Padman, and D. Kaplan, "On data reliability assessment in accounting information systems," *Inf. Syst. Res.*, vol. 16, no. 3, pp. 307–326, Sep. 2005, doi: 10.1287/isre.1050.0063.
44. Z. Abedjan, L. Golab, and F. Naumann, "Profiling relational data: a survey," *VLDB J.*, vol. 24, no. 4, pp. 557–581, Aug. 2015, doi: 10.1007/s00778-015-0389-y.
45. F. Naumann, "Data profiling revisited," *SIGMOD Rec.*, vol. 42, no. 4, pp. 40–49, Feb. 2014, doi: 10.1145/2590989.2590995.
46. F. Ridzuan and W. M. N. Wan Zainon, "A review on data cleansing methods for big data," *Procedia Comput. Sci.*, vol. 161, pp. 731–738, Jan. 2019, doi: 10.1016/j.procs.2019.11.177.
47. E. Rahm and H. H. Do, "Data cleaning: problems and current approaches," *IEEE Data Eng. Bull.*, vol. 23, no. 4, pp. 3–13, 2000.
48. A. Nambiar and D. Mundra, "An overview of data warehouse and data lake in modern enterprise data management," *Big Data Cogn. Comput.*, vol. 6, no. 4, p. 132, Dec. 2022, doi: 10.3390/bdcc6040132.
49. M. E. M. El Aissi *et al.*, "Data Lake Versus Data Warehouse Architecture: A Comparative Study," in *WITS 2020*, S. Bennani, Y. Lakhrissi, G. Khaissidi, A. Mansouri, and Y. Khamlichi, Eds., Singapore: Springer, 2022, pp. 201–210. doi: 10.1007/978-981-33-6893-4_19.

50. B. Guntupalli, "Data lake vs. data warehouse: choosing the right architecture," *Int. J. Artif. Intell. Data Sci. Mach. Learn.*, vol. 4, no. 4, pp. 54–64, Dec. 2023, doi: 10.63282/3050-9262.IJAIDSML-V4I4P107.
51. E. Begoli, I. Goethert, and K. Knight, "A lakehouse architecture for the management and analysis of heterogeneous data for biomedical research and mega-biobanks," in *2021 IEEE International Conference on Big Data (Big Data)*, Dec. 2021, pp. 4643–4651. doi: 10.1109/BigData52589.2021.9671534.
52. J. Eder and V. A. Shekhovtsov, "Data quality for federated medical data lakes," *Int. J. Web Inf. Syst.*, vol. 17, no. 5, pp. 407–426, Jul. 2021, doi: 10.1108/IJWIS-03-2021-0026.
53. D. E. O'Leary, "Embedding AI and crowdsourcing in the big data lake," *IEEE Intell. Syst.*, vol. 29, no. 5, pp. 70–73, Sep. 2014, doi: 10.1109/MIS.2014.82.
54. H. W. Kim, A. Mungroo, and J.-H. Lewe, "From data lake to digital twin: a data management framework for BEM and BAS interoperable buildings," *ASHRAE Trans.*, vol. 130, pp. 607–615, 2024.
55. A. Talwar, A. A. Talwar, R. B. Broach, L. H. Ungar, D. A. Hashimoto, and J. P. Fischer, "Artificial intelligence unlocks the healthcare data lake," *Artif. Intell. Surg.*, vol. 5, no. 2, pp. 239–246, May 2025, doi: 10.20517/ais.2024.109.

9 Future Perspectives on Process Analytical Technologies and Data Analytics for Biologics Manufacturing

Rupa Haldavnekar and Amine Kamen

9.1 INTRODUCTION

Biologics manufacturing has entered a period of profound transformation driven by digitalization, real-time analytics, and intelligent automation.[1–3] The rising therapeutic complexity of monoclonal antibodies, viral vectors, fusion proteins, recombinant vaccines, and mRNA modalities demands a manufacturing paradigm shift capable of anticipating variability rather than reacting to it.[4,5] This shift is redefining Process Analytical Technology (PAT) from a compliance-oriented monitoring framework into an integrated, predictive, and self-optimizing control system grounded in mechanistic modeling, multivariate analytics, and emerging digital-twin technologies.[1,6] Next-generation PAT will integrate diverse sensing modalities to capture physical, biochemical, and metabolic parameters simultaneously.[3] Advances in spectroscopy, dielectric sensing, and biosensor miniaturization are enabling multiplexed platforms capable of monitoring cell growth, nutrient flux, and product formation in real time.[4] Hybrid systems that couple vibrational spectroscopy (e.g., Raman, NIR) with chromatographic or mass spectrometric interfaces will provide orthogonal data streams and enhance sensitivity and robustness while reducing offline analysis.[5]

As biologics become more structurally and functionally complex, they exhibit heightened sensitivity to subtle variations in cell metabolism, raw-material attributes, microenvironmental stresses, and unit-operation dynamics.[7] Traditional control strategies—based on isolated measurements or retrospective assays—can no longer ensure the robustness, consistency, and regulatory confidence demanded by modern biopharmaceutical production.[5,8] Rapid advances in analytical technologies, including multi-attribute methods (MAM), rapid microbial testing, and in situ Raman spectroscopy, are enabling deeper attribute understanding and improved control strategies.[5,8] Yet global adoption remains challenged by inconsistent regulatory expectations and the absence of harmonized guidance for advanced data-driven systems.[9,10]

DOI: 10.1201/9781032662961-9

PAT has advanced significantly in recent years, moving toward adaptive control, hybrid mechanistic–machine-learning modeling, predictive maintenance, and closed-loop automation.[5,6,9] Digital twins—a virtual representation of a bioprocess that integrates first-principles models with real-time sensor data and artificial intelligence (AI)—play a central role in this evolution.[10,11] While digital twins are increasingly used for soft sensing, what-if simulations, and scale-up risk analysis,[12,13] no fully realized, end-to-end digital twin exists for biologics manufacturing. Limitations arise from incomplete biological knowledge, data sparsity, uncertainty quantification challenges, and the need for standardized model governance.

The COVID-19 pandemic accelerated these trends globally. As an example, Canada invested in a rapid development and deployment of mRNA vaccines through unprecedented investment in biomanufacturing infrastructure.[4,14] Federal initiatives such as Canada's Biomanufacturing and Life Sciences Strategy and the establishment of new vaccine manufacturing centers strengthened the national ecosystem for advanced bioprocessing. Canadian academic–industry collaborations have been especially targeted to capitalize on PAT innovations, including Raman-based real-time monitoring, chemometric modeling, microfluidic bioprocess intensification, and hybrid AI–mechanistic modeling for cell therapy and vaccine production. These efforts contributed to positioning Canada as a strategic player in scalable, digitally enabled biologics manufacturing.[15–20]

Regulatory agencies worldwide—including FDA, EMA, PMDA, and Health Canada—are increasingly supportive of advanced analytics and digital tools. Health Canada has emphasized flexibility and science-based decision-making, leveraging frameworks aligned with ICH Q8–Q14 and actively participating in global harmonization efforts. Through initiatives such as the Regulatory Innovation Sandbox and collaboration with the International Coalition of Medicines Regulatory Authorities (ICMRA), Health Canada has shown openness to model-informed drug development (MIDD), PAT-enabled control strategies, and real-time release (RTR) testing. However, challenges remain in establishing clear pathways for validating hybrid AI–mechanistic models, assessing digital-twin credibility, managing adaptive systems, and integrating cloud-based analytics under Good Manufacturing Practice (GMP) expectations. Canada's unique regulatory–innovation environment provides both an opportunity and a testbed for shaping next-generation oversight frameworks.[21–24]

This chapter provides a comprehensive and future-facing examination of PAT and data analytics in biologics manufacturing from a global perspective. Integrating industrial case studies and international regulatory developments, it outlines how real-time sensing, mechanistic and hybrid modeling, AI-driven analytics, and digital-twin frameworks will redefine bioprocess development, scale-up, and commercial production.[2,9,11] Through examples—ranging from digital twins for mRNA vaccine manufacturing to Raman-enabled adaptive control platforms and machine learning–guided design of gene therapy vectors—we articulate a roadmap for the next decade of biologics manufacturing, emphasizing predictable, resilient, and flexible production ecosystems supported by forward-looking regulatory environments.

9.2 ADVANCES IN PROCESS ANALYTICAL TECHNOLOGY: EXPANDING THE SENSING LANDSCAPE AND THE DEVELOPMENT OF PAT ANALYTICS

The evolution of PAT is inseparable from the rapid diversification and increasing complexity of modern biologics manufacturing. As production expands beyond monoclonal antibodies into cell and gene therapies, viral vectors, plasmid DNA, mRNA vaccines, lipid nanoparticles (LNPs), engineered exosomes, and emerging recombinant platforms, manufacturers require high-resolution, real-time, and modality-agnostic sensing systems capable of capturing biochemical, biophysical, and process-level dynamics with exceptional fidelity. To meet this need, contemporary PAT integrates a wide range of optical, spectroscopic, acoustic, electrochemical, and microfluidic tools that collectively provide continuous insight into chemical composition, biological activity, physical states, and process conditions.

9.2.1 Expanding the Sensing Toolbox

Among these modalities, Raman spectroscopy has emerged as one of the most transformative PAT technologies. Raman's chemically specific vibrational signatures allow simultaneous quantification of nutrients, metabolites, amino acids, impurities, and product-related attributes without labeling or sample preparation. Substantial advances in probe engineering, laser stability, spectral filtering, embedded chemometrics, and process-ready hardware have elevated Raman from an experimental instrument to a robust manufacturing tool. At commercial scale, organizations such as WuXi Biologics have implemented Raman-based PAT to monitor glucose, lactate, glutamine, ammonia, viable cell density, and early glycosylation trends—enabling adaptive feeding, enhanced process stability, and more consistent product quality. Achieving this level of robustness required extensive calibration, global model development, and harmonization across clones, scales, and instrument platforms.[25]

Complementing Raman, flow-cell, and microfluidic PAT devices—exemplified by technologies pioneered and patented by Nirrin Technologies—are reshaping the sensing landscape.[26] These aseptic, fouling-resistant architectures integrate directly with bioreactors and downstream steps, enabling high-frequency, continuous optical interrogation of conditioned media, intermediates, and purified streams. Flow-cell PAT supports real-time monitoring of protein aggregation, particle size distribution, impurity profiles, protein concentration, and spectral fingerprints under stable optical conditions. Critically, these systems capture transient events such as aggregation onset, shear-induced stress, and rapid metabolic shifts—phenomena often missed by low-frequency offline assays but highly consequential for product consistency and process robustness.

Beyond Raman and flow-cell technologies, the broader PAT landscape continues to expand through innovations in the following:

- Fluorescence-based PAT, now incorporating fluorescence lifetime spectroscopy and multi-parametric probes capable of reporting on protein folding

states, conformational microenvironments, redox status, and early aggregation signatures. When analyzed with machine-learning algorithms, these systems serve as highly sensitive early-warning indicators of structural or metabolic instability.
- Dielectric spectroscopy, which has evolved from a tool for viable cell density estimation to a sophisticated platform capable of detecting intracellular water mobility, membrane capacitance, apoptosis, metabolic shifts, and viral infection dynamics—particularly valuable in HEK cell cultures used for lentiviral, and adeno-associated virus (AAV) production systems.
- Acoustic and ultrasonic PAT, enabling inline monitoring of microbubble formation, viscosity changes, aggregation behavior, and shear-related mechanical effects.
- Microfluidic-integrated sensors, which increasingly incorporate digital polymerase chain reaction (PCR), affinity binding assays, or nanoparticle tracking analysis. These platforms could soon allow inline quantification of exosome concentrations, viral genome copies, LNP size distributions, or residual gene-editing reagents—measurements that currently rely on slow, offline assays.

As biologics become more modality-diverse and structurally complex, real-time impurity characterization is emerging as a critical frontier. Sensors capable of detecting host cell DNA, host cell proteins, lipid impurities from mRNA–LNP manufacturing, or residual CRISPR reagents for genome editing will be essential for robust purification. Embedding such tools directly into chromatography, ultrafiltration, or diafiltration operations may enable dynamic downstream processing where buffer conditions, flow rates, and step transitions are adjusted in real time based on impurity profiles rather than static recipes.

Together, the expansion of these sensing modalities is generating increasingly high-dimensional datasets, requiring equally sophisticated analytical methods to extract actionable process understanding. Embedded optical and electrochemical microsensors are transforming process monitoring by enabling in situ, non-invasive measurements.[6] These sensors can function across continuous bioprocess platforms, supporting distributed PAT networks and process intensification strategies.[7]

9.2.2 Hybrid PAT Models as the Dominant Future Approach

Hybrid models overcome the limitations of purely mechanistic or purely data-driven approaches by integrating physics-based foundations with the flexibility of machine learning (ML). As highlighted by Rathore and colleagues,[27] hybrid modeling is poised to become the dominant strategy for advanced PAT because it enhances model robustness, mitigates overfitting, ensures physical plausibility, enables extrapolation beyond the training domain, and supports real-time adaptive decision-making under biological variability. In these frameworks, dense PAT data streams—such as Raman spectra, dielectric properties, fluorescence lifetimes, and inline metabolite measurements—are mapped onto mechanistic parameters using neural networks, Gaussian process regression, or latent-variable architectures. This tight coupling allows continuous recalibration based on actual process behavior, removing the static assumptions that constrain classical mechanistic simulations and making hybrid PAT far better suited for dynamic, living bioprocesses.

9.2.3 Hybrid Soft Sensors as a Key Outcome

One of the most tangible industrial advances enabled by hybrid modeling is the emergence of **hybrid soft sensors**—virtual sensing tools that infer critical quality attributes (CQAs) or hidden process states by synchronizing sparse offline measurements, mechanistic equations, and high-frequency PAT signals. These systems blend the interpretability of first-principles models with the adaptability of ML, creating real-time estimators of process performances that would be impossible to measure directly. For example, a hybrid soft sensor may couple a mechanistic representation of cell growth, nutrient uptake, and metabolic fluxes with Raman-based glucose and amino-acid profiles, fluorescence indicators of protein folding or stress responses, and ML estimators for glycosylation trajectories or aggregation propensity.[27] Instead of relying on a single physical sensor, these virtual instruments compute CQAs through the integrated logic of the hybrid model, continuously recalibrated against live process data. Their industrial value is profound: hybrid soft sensors enable RTR testing, reduce the latency of identifying process deviations, support multivariate and model-predictive control strategies, and underpin dynamic optimization frameworks that adjust feeding, aeration, or mixing profiles in anticipation of biological variability.[28] As biologics processes grow in complexity and advanced manufacturing platforms—cell therapies, viral vectors, mRNA platforms—soft sensors provide a scalable path toward regulatory-aligned, data-dense, self-correcting manufacturing systems, positioning hybrid PAT as a foundational enabler of next-generation bioprocess automation.

9.3 DATA ANALYTICS IN THE DIGITAL BIOMANUFACTURING ERA: DEVELOPMENT OF PAT ANALYTICS; FROM CHEMOMETRICS TO ARTIFICIAL INTELLIGENCE

Chemometrics has long been central to PAT, using tools like principal component analysis (PCA) and partial least squares (PLS) for multivariate interpretation. However, as process data become more complex, non-linear, and time-resolved, AI-driven analytics are emerging as essential tools for process modeling and control. Deep learning, reinforcement learning, and hybrid AI frameworks can capture dynamic process relationships, enabling adaptive optimization and fault detection. The future development of PAT analytics will be defined by the convergence of mechanistic modeling and ML. Mechanistic models have long served as the scientific foundation for describing nutrient kinetics, metabolic fluxes, mass transfer, protein aggregation pathways, viral infection or replication kinetics, and physicochemical transformations. Their strength lies in interpretability, physical consistency, and the ability to extrapolate beyond observed operating ranges. However, purely mechanistic models encounter limitations when faced with biological complexity, non-linear interactions, unknown mechanisms, and context-dependent behaviors that violate simplified assumptions.

ML offers complementary strengths. ML can extract patterns, correlations, and functional relationships from high-dimensional PAT datasets—including Raman spectra, fluorescence signatures, dielectric responses, acoustic signals, and microfluidic sensor outputs—without requiring explicit a priori knowledge of underlying mechanisms. Future PAT analytics will rely on data fusion—integrating heterogeneous datasets (spectral, process, omics) into unified digital frameworks. Bayesian inference, kernel-based methods, and representation learning can correlate upstream and downstream variables, improving mechanistic understanding and prediction accuracy. Distributed computing architectures will define the future of PAT. Edge analytics enable low-latency decision-making, while cloud systems facilitate large-scale data integration, cross-site learning, and digital-twin synchronization. The combination enhances scalability and resilience in global manufacturing networks. Nevertheless, fully data-driven models lack interpretability, are sensitive to domain shift, and often require large, diverse datasets that are not always feasible in GMP settings.

9.3.1 Expanding the Scope of ML in Future PAT

The work from Stanford's Gao Lab further highlights how ML can explore vast biological design spaces.[4] Their deep learning approaches for protein and vector engineering—particularly for gene therapy—demonstrate that ML can map high-dimensional sequence–function landscapes inaccessible to hypothesis-driven experiments. As AAV, lentiviral vectors, and engineered enzymes become central to biopharmaceutical pipelines, models that link sequence attributes with manufacturability constraints and process behavior will guide the design of molecules compatible with robust, scalable production. This represents a fundamental shift in PAT analytics: ML will increasingly intersect with molecular design to ensure downstream manufacturability.

9.3.2 Digital Twins in Biologics Manufacturing: Promise, Progress, and Reality

Digital twins have rapidly evolved from their engineering origins—where real-time, physics-based simulations mirror aircraft, engines, or semiconductor systems—into powerful aspirational tools for the biopharmaceutical industry. In biologics manufacturing, digital twins are conceptualized as dynamic, continuously updating virtual representations of a bioprocess that integrate multiscale data streams, high-fidelity mechanistic models, and advanced analytics. Unlike traditional models used only during development, digital twins are intended to evolve alongside the physical process, assimilating streaming PAT data and adjusting predictions as new information becomes available.

Recent advances in PAT technologies, real-time sensing, mechanistic modeling, and ML have made digital twins increasingly feasible. Their impact is already evident in selected applications, including modeling steps in mRNA vaccine production,

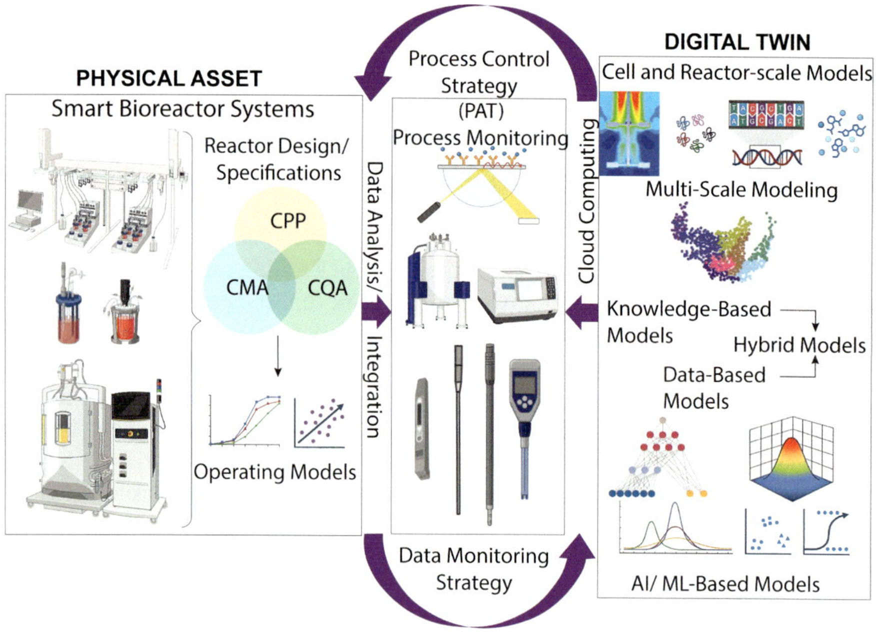

FIGURE 9.1 Digital twin concept and communication strategy with the physical asset.

optimizing adjuvant formulation, supporting facility-wide logistics simulations, and enabling operator training environments.[29–31] Hybrid digital twins (see Figure 9.1), which combine first-principles models with data-driven corrections, capture complex biological behaviors more accurately than purely mechanistic or purely data-driven approaches, providing predictive insight into metabolic shifts, glycosylation trajectories, mass-transfer limitations, or mixing nonuniformities that are difficult to measure directly.

In practice, digital twins function as continuously updating, high-resolution virtual representations of bioprocesses. Early implementations—such as those developed for mRNA vaccine production during the COVID-19 response—modeled enzymatic transcription dynamics, purification performance, and lipid nanoparticle (LNP) formulation parameters, revealing critical factors influencing yield, raw-material sensitivity, and technology transfer. Similarly, digital twins for vaccine adjuvant production demonstrated how fluid dynamics, particle interactions, and thermal distributions can be simulated to improve formulation consistency. At the system level, some contract development and manufacturing organizations now deploy multi-layer digital twins that integrate equipment scheduling, material flows, facility readiness, and compliance monitoring, enabling faster scenario evaluation and risk-informed planning.

Despite these advances, a fully functional, end-to-end digital twin for biologics manufacturing does not yet exist. Current implementations, while increasingly sophisticated, remain partial, domain-specific, and constrained by scientific, technological, and regulatory limitations.

First, the biological variability inherent to cell-based systems poses a fundamental challenge. Mechanistic models alone cannot capture cell-line–specific metabolic responses, nutrient limitations, or microenvironmental fluctuations, while data-driven models struggle with extrapolation, interpretability, and sparsity during early development. As a result, today's digital twins—though informed by advanced hybrid modeling—cannot yet deliver the complete process fidelity required for autonomous operation.

Second, real-time measurement capabilities remain insufficient. Digital twins depend on high-dimensional, continuous data streams, yet many CQAs are not directly measurable inline. Soft sensors can approximate unmeasured states such as glycosylation patterns, metabolic fluxes, or intracellular stress responses, but their reliability varies with process conditions, limiting the confidence needed for fully closed-loop control.

Third, the data ecosystem in biomanufacturing is fragmented. PAT instruments, historian databases, manufacturing execution systems (MES), and laboratory information management systems (LIMS) often operate in silos, lacking unified ontologies and standardized interfaces. Fully operational digital twins require seamless bidirectional communication between physical and virtual systems—something the current infrastructure only partially supports.

Finally, the regulatory landscape remains in transition. While agencies are developing frameworks for model lifecycle management, verification, and validation, digital twins are not yet accepted as autonomous control mechanisms or surrogates for end-product testing. Consequently, current twins function primarily as prediction and advisory tools rather than regulatory-grade engines of RTR or autonomous decision-making.

Collectively, these constraints explain why digital twins—despite significant progress—have not yet achieved their envisioned role as comprehensive, continuously learning representations of biologics manufacturing processes. Yet the trajectory is clear. The integration of high-resolution PAT data, mechanistic models, and ML is driving the emergence of hybrid digital twins that can interpret dynamic process behavior, detect early deviations, and support predictive control strategies. As advances in sensing, automation, and regulatory science continue, digital twins are poised to evolve from powerful analytical aids to foundational components of the next-generation biomanufacturing ecosystem.

9.4 CENTRAL ROLE OF PROCESS ANALYTICAL TECHNOLOGY (PAT) TO ESTABLISH ROBUST MANUFACTURING PLATFORMS FOR HEALTH EMERGENCY RESPONSES

Pandemic preparedness depends on the ability to rapidly develop, scale, and reliably manufacture vaccines and therapeutics under conditions of extreme urgency and uncertainty, including the emergence of future *disease X*. Initiatives such as the 100 Days Mission, endorsed by the G7 and advanced by the coalition for epidemic preparedness innovations (CEPI),[32] emphasize the need for platform-based, agile biomanufacturing systems capable of delivering safe and effective countermeasures at unprecedented speed. Within this context, ensuring robustness, consistency, and regulatory confidence in biologics manufacturing is not merely an operational goal but a critical component of national and global public health resilience.

PAT is foundational to achieving this level of preparedness. By enabling continuous, real-time monitoring of critical process parameters (CPPs) and CQAs, PAT shifts biologics manufacturing from a reactive, batch-centric paradigm toward a proactive, data-driven operating model. Advanced sensing modalities—including Raman and infrared spectroscopy, chromatography, multi-angle light scattering, and inline optical and electrochemical sensors—provide high-frequency visibility into complex biological processes that were previously assessed only through end-point testing. This real-time insight enables earlier detection of deviations, reduces batch failures, accelerates development timelines, and enhances process reproducibility, all of which are essential during pandemic-scale production.

A key contribution of PAT to pandemic readiness is its role in RTR and pre-approval of processes (PAP). RTR leverages continuous PAT data to verify that each batch meets predefined quality criteria without waiting for traditional post-production testing, thereby accelerating product availability while maintaining safety and efficacy standards. PAP complements this approach by using continuous monitoring and predictive modeling to demonstrate process robustness and reproducibility prior to full-scale deployment, streamlining regulatory review and enabling rapid technology transfer across sites. Together, RTR and PAP allow speed and regulatory rigor to coexist—an essential requirement during emergency response scenarios.[33,34]

The COVID-19 pandemic demonstrated the practical value of PAT-enabled manufacturing. In mRNA vaccine production, real-time monitoring of enzymatic transcription, purification efficiency, and LNP formulation allowed manufacturers to dynamically adjust operating conditions, ensuring high yield, consistency, and product quality despite unprecedented timelines and global demand. Beyond immediate control, PAT-supported predictive modeling enabled risk-based decision-making, rapid scale-up, and scenario planning, strengthening supply chain resilience during periods of acute stress.

Increasingly, PAT data serve as the backbone for hybrid and emerging digital-twin frameworks in biomanufacturing. By integrating mechanistic models with real-time sensor data and ML, digital twins provide a system-level view of manufacturing operations, enabling simulation of scale-up strategies, material flow adjustments, and facility readiness under emergency conditions. These virtual representations help identify bottlenecks, optimize resource allocation, and coordinate upstream and downstream operations, enhancing both agility and reliability across distributed manufacturing networks.

From a regulatory perspective, PAT aligns manufacturing innovation with compliance by ensuring that automated decisions remain transparent, traceable, and auditable. Cloud-enabled analytics, unified data architectures, and continuous process verification support regulatory expectations for control, documentation, and lifecycle management, even as manufacturing systems become more autonomous.

By reducing reliance on manual intervention and retrospective testing, PAT minimizes human error while strengthening quality assurance.

In summary, PAT is not simply an analytical add-on but the foundation of reliable, regulation-aligned, and pandemic-ready biomanufacturing. Its integration with RTR, PAP, predictive modeling, and digital-twin technologies enables faster, safer, and more resilient responses to public health crises. As biologics manufacturing moves toward globally distributed, digitally orchestrated platforms capable of producing multiple therapeutic modalities, PAT will remain central to ensuring that speed, quality, and regulatory confidence advance together.

9.5 CONCLUDING REMARKS

The ongoing digital transformation of biologics manufacturing is reshaping how processes are monitored, controlled, and validated. Hybrid models—integrating mechanistic insight with machine-learning flexibility—are emerging as the foundational architecture for next-generation PAT, enabling soft sensors, predictive control, and RTR strategies that were previously unattainable. These advances are not simply incremental improvements: they redefine the entire manufacturing paradigm by shifting from reactive, batch-dependent quality assurance to proactive, data-dense, continuously self-optimizing operations. Yet despite impressive progress, the field remains in transition. No end-to-end digital twin with feedback control loops capable of fully virtualizing and controlling a commercial biologics facility exists today, and major challenges persist in data standardization, model transferability, regulatory acceptance, and the representation of biological variability. Nevertheless, the trajectory is clear. As sensing technologies mature, algorithms become more explainable, cloud architectures integrate seamlessly with automation, and regulators increasingly support data-centric quality frameworks; hybrid PAT systems will catalyze a new era of resilient, agile, and globally scalable biomanufacturing. Ultimately, the convergence of PAT, hybrid modeling, and digital twins will form the backbone of future therapeutic production—enabling faster responses to emerging health threats, more consistent product quality, and fundamentally more intelligent manufacturing ecosystems.

REFERENCES

1. Schmidt, A., Helgers, H., Vetter, F.L., Juckers, A., Strube, J. Digital twin of mRNA-based SARS-COVID-19 vaccine manufacturing towards autonomous operation for improvements in speed, scale, robustness, flexibility and real-time release testing. *Processes* 9, 748 (2021). https://doi.org/10.3390/pr9050748.
2. Harutyunyan, R. Case Study: How Pfizer Used Meta Quest and Digital Twins to Accelerate Global Vaccine Production. *Knoxlabs*. https://www.knoxlabs.com/blogs/vr-xr-news/case-study-how-pfizer-used-meta-quest-and-digital-twins-to-accelerate-global-vaccine-production.
3. SAMSUNG BIOLOGICS. Accelerating Biomanufacturing through Integrated Digital Twin Technologies. https://www.fiercepharma.com/sponsored/accelerating-biomanufacturing-through-integrated-digital-twin-technologies (2025).

4. Gao, W., Mahajan, S. P., Sulam, J. & Gray, J. J. Deep learning in protein structural modeling and design. *Patterns* **1**, 100142 (2020).
5. Rathore, A.S., Mishra, S., Nikita, S., Priyanka, P. Bioprocess control: Current progress and future perspectives. *Life 11*, 557 (2021). https://doi.org/10.3390/life11060557
6. ©BioPhorum Operations Group Ltd. Defining Digital Twins in a Biomanufacturing Environment. https://www.biophorum.com/download/defining-digital-twins-in-a-biomanufacturing-environment-biophorum/ (2025).
7. Zhou, H. WuXi Biologics Stands on PAT. *GEN – Genetic Engineering and Biotechnology News.* https://www.genengnews.com/resources/tutorials/wuxi-biologics-stands-on-pat/ (2025).
8. Wang, T. *et al.* Advancing the implementation of innovative analytical technologies in pharmaceutical manufacturing—Some regulatory considerations. *J. Pharm. Sci.* **114**, 816–828 (2025).
9. Helgers, H., Hengelbrock, A., Schmidt, A., Strube, J. Digital Twins for Continuous mRNA Production. *Processes* **9**, 1967 (2021). https://doi.org/10.3390/pr9111967
10. Göppert, A. *et al.* Pipeline for ontology-based modeling and automated deployment of digital twins for planning and control of manufacturing systems. *J. Intell. Manuf.* **34**, 2133–2152 (2023).
11. Brühlmann, D. Bioprocess 4.0 & Digital Twins: Why Integrated Continuous Biomanufacturing Is Your Next Strategic Move. *Brühlmann Consulting.* https://bruehlmann-consulting.com/data-analytics/bioprocess-4-0-digital-twins-why-integrated-continuous-biomanufacturing-is-your-next-strategic-move/ (2025).
12. Cao, H. *et al.* Towards AI-Based Sustainable and XR-Based Human-Centric Manufacturing: Implementation of ISO 23247 for Digital Twins of Production Systems. Preprint at https://doi.org/10.48550/arXiv.2508.14580 (2025).
13. Xu, W. & Xie, W. RAPTOR-GEN: RApid PosTeriOR GENerator for Bayesian Learning in Biomanufacturing. Preprint at https://doi.org/10.48550/arXiv.2509.20753 (2025).
14. Mehta, U. Implementation of Raman Spectroscopy for in-line monitoring of critical process parameters of CHO cell perfusion cultures. *Integr. Contin. Biomanufacturing VI.* https://dc.engconfintl.org/biomanufact_vi/38 (2024).
15. adMare BIOINNOVATION. New adMare Institute Report Examines Capital Flows in Canadian Life Sciences and the Urgent Need to Capture Domestic Economic Value. July 09 2025 https://www.admarebio.com/en/news-details/new-admare-institute-report-examines-capital-flows-in-canadian-life-sciences.
16. Government of Canada. Biologics Manufacturing Centre. https://nrc.canada.ca/en/research-development/nrc-facilities/biologics-manufacturing-centre (2020).
17. Bridge S. and Roumeliotis I. Canada building facilities to make vaccines for COVID-19 and other viruses. *CBC News* (January 12, 2021). https://www.cbc.ca/news/health/vaccine-development-production-in-canada-1.5868398
18. Entos Pharmaceuticals. $198.5 M GoC&GoA Partnership to Biomanufacture Next Gen Medicines. https://entospharma.com/news-%26-media/f/1985-m-gocgoa-partnership-to-biomanufacture-next-gen-medicines (2025).
19. Overview of Canada's Biomanufacturing and Life Sciences Strategy. https://ised-isde.canada.ca/site/ised/en/programs-and-initiatives/health-emergency-readiness-canada/overview-canadas-biomanufacturing-and-life-sciences-strategy (2025).
20. Government of Canada. Biomanufacturing: Advancing Life Sciences Industry in Canada – Home. https://ised-isde.canada.ca/site/ised/en/programs-and-initiatives/health-emergency-readiness-canada (2023).
21. BIOTECanada. Modernize NSNR. https://www.biotech.ca/news/modernize-nsnr/ (2024).
22. Department of Finance Canada. Chapter 4 (archived): Economic Growth for Every Generation, Budget 2024. https://budget.canada.ca/2024/report-rapport/chap4-en.html (2024).

23. Innovation, Science and Economic Development Canada. Government of Canada Announces Major Milestone in the Canadian Biomanufacturing Sector. https://www.canada.ca/en/innovation-science-economic-development/news/2025/09/government-of-canada-announces-major-milestone-in-the-canadian-biomanufacturing-sector.html (2025).
24. Canadian News Wire (Cision). Canada supports biomanufacturing capacity to advance next generation medicines. *Yahoo Finance.* https://ca.finance.yahoo.com/news/canada-supports-biomanufacturing-capacity-advance-174100192.html (2025).
25. Chen, G., Hu, J., Qin, Y. & Zhou, W. Viable cell density on-line auto-control in perfusion cell culture aided by *in-situ* Raman spectroscopy. *Biochem. Eng. J.* **172**, 108063 (2021).
26. Corsino, J. Nirrin Technologies Granted Patent for Cutting-Edge Flow-Cell Technology, Advancing Real-Time Bioprocess Monitoring. *Nirrin.* https://www.nirrin.tech/nirrin-technologies-granted-patent-for-cutting-edge-flow-cell-technology-advancing-real-time-bioprocess-monitoring/ (2025).
27. Rathore, A. S. & Winkle, H. Quality by design for biopharmaceuticals. *Nat. Biotechnol.* **27**, 26–34 (2009).
28. US Food & Drug Administration. Sterile Drug Products Produced by Aseptic Processing — Current Good Manufacturing Practice. https://www.fda.gov/regulatory-information/search-fda-guidance-documents/sterile-drug-products-produced-aseptic-processing-current-good-manufacturing-practice (2020).
29. Phalak, P., *et al.* Digital twin implementation for manufacturing of adjuvants. *Processes 11*, 1717 (2023). https://doi.org/10.3390/pr11061717.
30. Nair A., Kis, Z. How Advanced PAT Aids Quality by Digital Design in mRNA Manufacturing. *Advancing RNA.* https://www.advancingrna.com/doc/how-advanced-pat-aids-quality-by-digital-design-in-mrna-manufacturing-0001 (2024).
31. Gerzon, G., Sheng, Y. & Kirkitadze, M. Process analytical technologies – Advances in bioprocess integration and future perspectives. *J. Pharm. Biomed. Anal.* **207**, 114379 (2022).
32. Delivering Pandemic Vaccines in 100 Days. What Will It Take? https://static.cepi.net/downloads/2024-02/CEPI-100-Days-Report-Digital-Version_29-11-22.pdf (2022).
33. Speder, B. & Dalemans, W. The Use of Prior Knowledge and Platform Approaches in Early Vaccine Development. *Regulatory Rapporteur.* https://www.regulatoryrapporteur.org/clinical-trials/the-use-of-prior-knowledge-and-platform-approaches-in-early-vaccine-development/960.article (2025).
34. Abdin, A.Y., De Pretis, F., Landes, J. Fast methods for drug approval: research perspectives for pandemic preparedness. *Int. J. Environ. Res. Public Health 20*, 2404 (2023). https://www.mdpi.com/1660-4601/20/3/2404.

10 Case Study
PAT in mRNA–LNP Manufacturing

Laurence Dubeau-Mallette, Julia Puppin Chaves Fulber and Amine A. Kamen

10.1 INTRODUCTION

Messenger RNA (mRNA) therapeutics are biological drugs that gained traction during the COVID-19 pandemic due to the billions of vaccine doses successfully administered (Krammer, 2024; Pardi & Krammer, 2024), establishing the mRNA modality as a primary vaccinology platform for pandemic preparedness with fast development and manufacturing times. mRNA is a versatile technology, as these drug products can encode proteins for various prophylactic and therapeutic applications including enzymes, self-amplifying viral components and antigens (Parhiz et al., 2024). Approved mRNA drugs currently include COVID-19 and RSV vaccines (American Society of Gene and Cell Therapy, 2025), with over 70 clinical trials ongoing as of June 2025 to expand the possible uses of this platform (Żak & Zangi, 2025).

While mRNA manufacturing is cell-free, the template DNA required for transcription can be produced through bacterial fermentation or enzymatic synthesis (Hersch et al., 2024; Schmidt et al., 2021). Since plasmid DNA (pDNA) manufacturing has been thoroughly discussed elsewhere (Agbogbo et al., 2023; Shankar et al., 2024), this chapter will focus on the steps of mRNA manufacturing, starting from a previously amplified, linearized and purified DNA template.

Production of mRNA begins with *in vitro* transcription (IVT) (Figure 10.1), an enzymatic reaction in which an RNA polymerase (RNAP) catalyzes the incorporation of nucleotides (nucleotide triphosphates (NTPs)) to form an mRNA molecule according to the template DNA sequence. Aside from encoding the therapeutic protein and untranslated regions (UTRs), mRNA must also include a 5′ cap and 3′ poly(A) tail to ensure efficient translation and stability. The 5′ cap can be introduced co-transcriptionally with incorporation of cap analogs by the RNAP or added post-transcriptionally in a separate enzymatic step (Beverly et al., 2016). Similarly, the poly(A) can be added post-transcriptionally through polyadenylation enzymes or co-transcriptionally by encoding it in the DNA template (Youssef et al., 2023). Following IVT, mRNA is purified through a chromatographic step, with additional filtration steps to exchange the buffer and/or concentrate the product as needed (Cui et al., 2023). The mRNA is then encapsulated into lipid nanoparticles (LNPs) through microfluidic or jet-based mixing with 4 lipids (Ahl, 2025). Last, the mRNA–LNPs undergo buffer exchange and sterile filtration to achieve the final formulation, possibly including freeze-drying to maximize shelf-life (Kafle et al., 2024).

 DOI: 10.1201/9781032662961-10

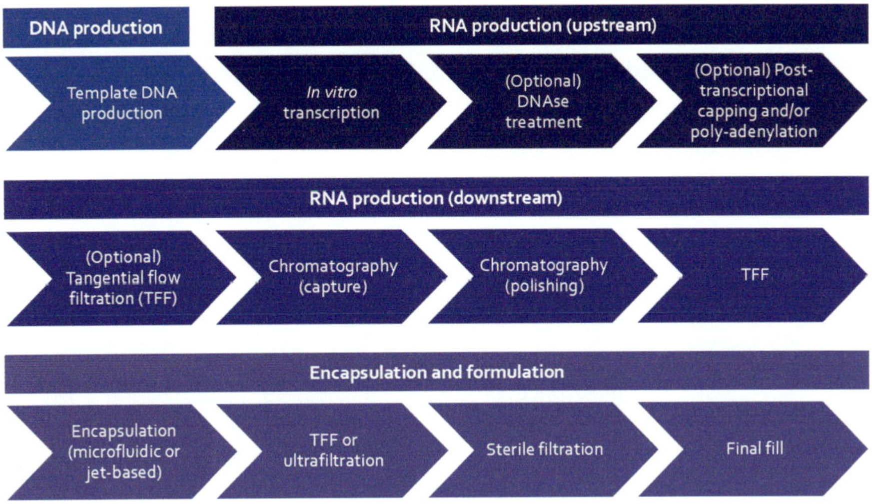

FIGURE 10.1 Schematic of each step in the mRNA manufacturing process including enzymatic synthesis, purification, encapsulation and final formulation of mRNA molecules. This is a general overview of a typical process, but individual steps may vary for each manufacturer or mRNA drug product.

As with other biologics, applying a quality by design (QbD) framework for mRNA manufacturing is crucial for ensuring robust processes and consistent product quality (Daniel et al., 2022; Guimaraes et al., 2024; Hu et al., 2024). Quality attributes of mRNA drug substances must be measured at multiple stages of the manufacturing process to enable effective process control and product consistency. Rather than relying only on end-product testing, process analytical technologies (PAT) emphasizes real-time or near-real-time assessment of critical quality attributes (CQAs) to support the understanding of process performance. mRNA–LNPs drugs present unique challenges compared to other therapeutic modalities like viral vectors or recombinant proteins, including specific impurities and associated analytical challenges. Following approval of mRNA-based COVID-19 vaccines, organizations such as US Pharmacopeia (USP) (United States Pharmacopeia, 2024) and the European Medicines Agency (European Medicines Agency, 2025) have issued guidelines and recommendations for quality control of mRNA products. This chapter provides an overview of the analytical tools currently used to evaluate key mRNA CQAs across upstream, downstream and formulation steps, and how these PAT-aligned analytics allow control of mRNA quality throughout the workflow.

10.2 PROCESS ANALYTICAL TECHNOLOGIES FOR ASSESSING mRNA QUALITY ATTRIBUTES

This section details well-established and novel techniques according to published guidelines (European Medicines Agency, 2025; United States Pharmacopeia, 2024) and recent literature, as summarized in Table 10.1.

TABLE 10.1
Analytical Tools Used for Measurement of Each Quality Attribute in mRNA–LNP Manufacturing

	Quality Attribute	Analytical Tools	Reference
	RNA concentration	UV spectroscopy	Green and Sambrook (2020)
		Fluorescence-based assay	Green and Sambrook (2020)
		HPLC	Barros et al. (2025), Korenč et al. (2021), Pregeljc et al. (2023), Welbourne et al. (2024)
		RT-qPCR/RT-ddPCR	Hu et al. (2024)
		CGE	Barros et al. (2025)
	RNA integrity and length	CGE	Barros et al. (2025), Camperi et al. (2024)
		Sequencing (nanopore)	Gunter et al. (2023)
		IP-RP-HPLC	Camperi et al. (2024), Packer et al. (2021)
		AEX HPLC	Webb et al. (2025)
		SEC	Webb et al. (2025)
	RNA sequence identity	Sequencing (high-throughput, sanger, nanopore)	Guimaraes et al. (2024), Hu et al. (2024)
		LC/MS	Jiang et al. (2019), Vanhinsbergh et al. (2022), Wolf et al. (2022)
RNA purity	**Presence of dsRNA species**	ELISA	Holland et al. (2024)
		Dot-blot	Baiersdörfer et al. (2019), Karikó et al. (2011), Piao et al. (2022)
		IP-RP-HPLC	Nwokeoji et al. (2017), Sousa Rosa et al. (2024)
		Electrophoresis	Peña et al. (2023, 2024)
	Presence of residual host cell DNA	qPCR/dPCR	Food and Drug Administration (2021), Lee et al. (2010), Li et al. (2021), Peper et al. (2014), Tian et al. (2024)
		Fluorescence (PicoGreen)	Lauro et al. (2022)
		Sequencing	Gunter et al. (2023)
	Residual template DNA	AEX HPLC	Leban et al. (2024)
		qPCR/dPCR	Leban et al. (2024)
	Residual T7 RNAP	ELISA	Shieh et al. (2024)
		Simple Western	Shieh et al. (2024)
	Residual proteins	NanoOrange	Madigan(2021)
		Silver staining	Madigan(2021)
	Presence of endotoxins (for safety)	Gel-clot technique	United States Pharmacopeia (2012)
		Turbidimetric technique	United States Pharmacopeia (2012)
		Chromogenic technique	United States Pharmacopeia (2012)
	5′ capping efficiency	LC/MS	Addepalli et al. (2025), Beverly et al. (2016), Cui et al. (2025), Galloway et al. (2020), Gau et al. (2023), Menneteau et al. (2025), Muthmann et al. (2022), Nakayama et al. (2023), Nguyen et al. (2021), Vlatkovic et al. (2022), Wang et al. (2025), Wolf et al. (2022)
		Ligation-mediated qPCR	Blewett et al. (2011), Ho (2017)
		Enzymatic degradation assay	Chiron and Jais 2017); Pregeljc et al. (2023)
	Poly(A) tail length	LC/MS	Beverly et al. (2018a), Gilar et al. (2023), Wang et al. (2025)
		CGE	Di Grandi et al. (2023), Su et al. (2025)
		IP-RP LC	Gilar et al. (2023)
		dPCR	Food and Drug Administration (2021)
		HILIC	Goyon et al. (2025)
		Sequencing	Chang et al. (2014), Garalde et al. (2018)

(Continued)

TABLE 10.1 (*Continued*)
Analytical Tools Used for Measurement of Each Quality Attribute in mRNA–LNP Manufacturing

Quality Attribute	Analytical Tools	Reference
RNA encapsulation efficiency	Fluorescence-based assay (RiboGreen)	McKenzie et al. (2023)
	AEX HPLC	Hara et al. (2024)
	RT-qPCR	Nag et al. (2022)
	Nano-flow cytometry (NanoFCM)	Nourafkan et al. (2025)
	Two-dimensional chromatography	Pavlin et al. (2025)
	UV–Vis	Galchenko and Böttger (2025), Gao et al. (2024)
Lipid identity	LC/MS	Fekete et al. (2023), Packer et al. (2021), Parot et al. (2024)
	RP-HPLC-CAD or RP-HPLC-ELSD	Mousli et al. (2022)
	IP-RP-HPLC (lipid adducts)	Packer et al. (2021)
LNP surface charge	DLS	McKenzie et al. (2023), Nourafkan et al. (2025)
	NTA	Parot et al. (2024)
	MALS	Gao et al. (2024), Jia et al. (2021), Pavlin et al. (2025)
LNP polydispersity	DLS	McKenzie et al. (2023), Nourafkan et al. (2025)
	NTA	Parot et al. (2024)
	MALS	Gao et al. (2024), Jia et al. (2021), Pavlin et al. (2025)
	SR-DLS	Nourafkan et al. (2025)
	NanoFCM	Nourafkan et al. (2025)
	ILM	Duffrène et al. (2025)
LNP size	DLS	McKenzie et al. (2023), Nourafkan et al. (2025)
	NTA	Parot et al. (2024)
	MALS	Gao et al. (2024), Jia et al. (2021), Pavlin et al. (2025)
	Electron microscopy	Di et al. (2022), Jia et al. (2021)
	SR-DLS	Nourafkan et al. (2025)
	NanoFCM	Nourafkan et al. (2025)
	ILM	Duffrène et al. (2025)
Potency	Cell-based assay	Cao et al. (2025), Feller et al. (2025), Li et al. (2023), Patel et al. (2023a)

HPLC: high-performance liquid chromatography; **RT:** reverse transcription; **qPCR:** quantitative polymerase chain reaction; **ddPCR:** digital droplet polymerase chain reaction; **IP-RP:** ion-pair reversed-phase; **AEX:** anion exchange; **SEC:** size-exclusion chromatography; **LC/MS:** liquid chromatography mass spectrometry; **CE:** capillary gel electrophoresis; **HILIC:** hydrophilic interaction liquid chromatography; **CAD:** charged aerosol detector; **ELSD:** evaporative light scattering detector; **DLS:** dynamic light scattering; **NTA:** nanoparticle tracking assay; **MALS:** multi-angle light scattering; **NanoFCM:** nano-flow cytometry; ILM: interferometric light microscopy; **SR-DLS:** spatially resolved dynamic light scattering.

10.2.1 RNA Concentration

RNA concentration is a critical attribute that should be evaluated throughout the manufacturing process to ensure adequate IVT yields, purification recovery, dosage of the final product and stability over time. There are a variety of methods available for quantification of either total RNA, RNA with a specific length or RNA with a specific sequence. For analysis of encapsulated RNA, refer to section "RNA encapsulation efficiency".

Total RNA can be determined based on ultraviolet (UV) absorbance measurements without any standard curves required, either through offline measurements on instruments like NanoDrop or inline detectors commonly used in chromatography systems. Since other components such as salts, solvents, proteins, free nucleotides and DNA impact UV absorbance, direct measurements on NanoDrop are unreliable for crude RNA samples and are, thus, more useful for quantification of purified mRNA (Green & Sambrook, 2020). Separation of RNA from other molecules through high-pressure liquid chromatography (HPLC), however, enables the use of UV detectors for accurate quantification of non-pure samples (Pregeljc et al., 2023; Welbourne et al., 2024). Fluorescent dye-based assays such as RiboGreen, on the other hand, can easily be used with non-pure RNA samples due to its compatibility with a variety of salts, although it will also detect DNA (Green & Sambrook, 2020). These methods will measure all RNA present in the sample, including shorter byproducts and potentially mutated RNA.

Many methods cited in the "RNA integrity & length" section can also be used for selective quantification of RNA molecules with a specific length. Ion-pair reversed-phase HPLC (IP-RP-HPLC) and capillary electrophoresis separate the sample by length into different peaks and concentration of RNA can be calculated by comparing the area under the curve of the target peak to a standard curve (Barros et al., 2025).

Sequence-specific RNA quantification can be performed through Reverse transcription-quantitative PCR (RT-qPCR) or Reverse Transcription Droplet Digital PCR (RT-ddPCR), in which primers are designed for the target sequence (Hu et al., 2024). Alternatively, oligo-dT HPLC can be used to quantify only mRNA containing a poly(A) tail, which excludes shorter RNA fragments or degraded products (Korenč et al., 2021).

10.2.2 RNA Integrity and Length

It is essential to evaluate RNA length to ensure presence of the full-length target RNA sequence and absence of smaller aborted transcripts and degraded mRNA or larger run-off transcripts, aggregates and double-stranded RNA (dsRNA) (Webb et al., 2025). Often, RNA integrity is reported as a percentage of the intact mRNA at the target length compared to all RNA of other lengths detected.

Although traditional agarose gel electrophoresis can be used, capillary gel electrophoresis (CGE) offers a quantitative read-out with higher throughput, resolution and sensitivity, making it more common for assessment of mRNA drugs in industry (Mantri et al., 2024; Webb et al., 2025). Both methods typically use denaturing conditions and apply an electric field to separate the negatively charged mRNA solely by length and eliminate the influence of secondary structures. An intercalating dye is used to detect the RNA by fluorescence (Camperi et al., 2024).

A variety of liquid chromatography (LC) methods coupled with UV detectors can also be used to characterize RNA integrity (Webb et al., 2025). IP-RP-HPLC relies on a hydrophobic stationary phase and a mix of solvents and ion-pairing reagents in the mobile phase. The ion-pairing salts neutralize the charged RNA backbone and increase hydrophobic interactions to allow for separation based on the RNA length and the hydrophobicity of aromatic nucleobases, meaning the RNA sequence influences separation in IP-RP-HPLC more than in CGE (Packer et al., 2021). Anion-exchange chromatography (AEX HPLC), on the other hand, uses a positively charged stationary phase to separate RNA molecules based on the length of the negatively charged backbone, but current methods have lower resolution than IP-RP-HPLC or CGE (Webb et al., 2025). Size exclusion chromatography (SEC) has also been investigated for mRNA and mRNA–LNPs, separating them by hydrodynamic radius based on the size of the pores in the stationary phase, although the focus with SEC has been on analyzing aggregation rather than RNA integrity (Webb et al., 2025).

RP-IP-HPLC and CGE are the most common methods to evaluate RNA integrity and have been compared in the literature (Barros et al., 2025; Camperi et al., 2024; Packer et al., 2021). RP-HPLC has been shown to detect impurities like lipid-mRNA adducts that were absent in CGE (Packer et al., 2021), while CGE has higher resolution when detecting RNA degradation (Barros et al., 2025). Although RP-HPLC might be more a more versatile approach for complex samples with different molecule types, CGE has much shorter analysis times, does not rely on solvents and migration times are more directly correlated to mRNA length (Barros et al., 2025). Ultimately, both methods are recommended to characterize RNA integrity (European Medicines Agency, 2025).

Long-read nanopore sequencing also enables assessment of RNA integrity, as each individual molecule is fully sequenced from start to finish (Gunter et al., 2023). Aside from determining the percentage of full-length product, this allows further investigation into the sequence of the abnormal transcripts to potentially identify specific issues related to the process or sequence design.

10.2.3 RNA Identity

mRNA identity is a CQA of mRNA vaccines and therapeutics because it is directly linked to drug efficacy. It is therefore essential to verify the mRNA sequence, as this would be a confirmation that the intended product has been correctly manufactured (Webb et al., 2025).

Sequencing methods are the most common tools for confirming RNA identity. Sanger sequencing, a cost-effective and rapid method, works by first reverse-transcribing the mRNA into complementary DNA (cDNA) using RT-PCR, then the resulting cDNA is sequenced using fluorescently labeled dideoxynucleotides (ddNTPs) that terminate DNA synthesis at specific bases (Guimaraes et al., 2024). For applications requiring higher throughput or broader sequence coverage, next-generation sequencing (NGS) is used (Guimaraes et al., 2024). This method involves fragmenting the RNA, making cDNA, ligating adapters, amplifying a library and sequencing in parallel (Guimaraes et al., 2024). As both methods rely on reverse transcription, which can be challenging when analyzing mRNA containing modified nucleotides, the use of nanopore sequencing is a suitable alternative: it allows direct sequencing of full-length native RNA molecules (Guimaraes et al., 2024). Some limitations are still being addressed regarding the workflow; however, as it evolves further it is expected to become an informative approach to RNA identity testing (Gunter et al., 2023).

Another approach that does not require conversion to cDNA, making it a direct approach, is mRNA sequence mapping by LC–MS (Webb et al., 2025). In this case, the mRNA is enzymatically digested by one or more RNases to generate smaller oligoribonucleotide fragments, which are subsequently separated by LC and analyzed using high-resolution tandem mass spectrometry (Webb et al., 2025). Enzymes used include RNase T1 and RNase 4 (Jiang et al., 2019; Vanhinsbergh et al., 2022; Wolf et al., 2022). Data is then analyzed to correlate the experimentally detected oligoribonucleotides, which are unique sequences that map only once to the mRNA, with the theoretical mRNA sequence (Webb et al., 2025).

10.2.4 RNA Purity

Ensuring the purity of mRNA is critical for product quality, efficacy and safety (Webb et al., 2025). During transcription, purification and formulation, a variety of contaminants can be introduced or carried over, each with potential impacts on potency, stability, innate immune activation or patient safety (Hu et al., 2024). Key analytes that must be monitored include residual host cell RNA, DNA and proteins, endotoxins and process-related impurities such as enzymes, leftover NTPs, remaining plasmid template DNA and dsRNA (Hu et al., 2024). A range of analytical tools is therefore required to assess content and purity throughout the manufacturing process.

10.2.4.1 Residual Host Cell DNA

Residual host cell DNA consists of nucleic acids originating from the production host (*Escherichia coli*) that remain after cell lysis, extraction and purification steps. These impurities can persist as fragments or intact molecules and may co-purify with the mRNA drug substance. Since they can present risks of infectivity, oncogenicity, immunogenicity and/or mutagenesis, their presence must be monitored throughout manufacturing and in the final product to ensure quality (European Medicines Agency, 2025; Food and Drug Administration, 2021).

Several analytical methods are available for their quantification. Quantitative PCR (qPCR) with primers specific to the template sequence is commonly used due to its high sensitivity. This method is established and widely accepted for detecting residual host cell DNA as it has been routinely applied in regulatory analyses across a wide range of therapeutic manufacturing processes that also involve fermentation in cells, including *E. coli* (Lee et al., 2010; Li et al., 2021; Peper et al., 2014). Droplet digital PCR (ddPCR works similarly but provides absolute quantification without the need for a standard curve and offers improved precision at low copy numbers (Food and Drug Administration, 2021; Tian et al., 2024). Different targets can be used for detecting *E. coli* genomic DNA, including 16S rRNA and 23S rRNA gene sequences, both present across the genome thus making them reliable proxies (Lee et al., 2010; Li et al., 2021; Peper et al., 2014; Tian et al., 2024).

Fluorescence-based assays, notably PicoGreen, provide a sensitive and rapid method for measuring total double-stranded DNA, although they do not distinguish between template DNA and other dsDNA species, meaning they are useful for measuring total DNA (Lauro et al., 2022).

More recently, Oxford Nanopore sequencing has also been applied to detect residual host cell DNA, providing a high-resolution method for identifying remaining genomic fragments, without the need for specific primers (Gunter et al., 2023).

10.2.4.2 Residual T7 RNAP

Following purification of the mRNA, it is necessary to ensure low levels of residual proteins in mRNA drug substances as they can trigger unintended immune responses (World Health Organization, 2021). However, T7 RNAP is particularly difficult to eliminate because of its strong affinity for nucleic acids. For this reason, its residual level is often used as a proxy for total residual protein content, quantified through a T7 RNAP-specific ELISA (Shieh et al., 2024). More recently, a new method was developed, consisting of automated capillary-based immunoassays, also known as Simple Western, which enables sensitive and reproducible quantification of residual T7 RNAP (Shieh et al., 2024).

10.2.4.3 Residual Host Cell Proteins

Although purification methods aim to remove as much as possible any contaminants, residual host cell proteins, coming from *E. coli*, or process-related proteins, such as pyrophosphatase, DNAse, proteinase K, cap analogs and poly(A) polymerase, can remain (World Health Organization, 2021). As mentioned before, since these can cause immune responses in patients, they must be measured to make sure they stay below a certain limit. Common methods include the NanoOrange assay and silver staining, both sensitive methods, but with a preference for NanoOrange when it comes to final product release because silver staining offers only qualitative results (Madigan, 2021).

10.2.4.4 Residual Template DNA

Residual template DNA refers to linearized pDNA that was used as the transcription template during IVT. Because this DNA can persist through downstream purification, its removal and accurate quantification are critical (Leban et al., 2024).

Similarly to measuring residual host cell RNA/DNA, residual pDNA can be measured via qPCR or ddPCR, with the only difference being the primers that target the template DNA sequence instead (Leban et al., 2024).

Chromatographic approaches are also being explored for the detection of linear pDNA. In one method, anion-exchange HPLC (AEX HPLC) coupled with prior mRNA digestion with RNase A allows for the separation of mRNA fragments from residual linear pDNA, and thus the quantification of the latter (Leban et al., 2024).

Contrary to PCR methods, it can be applied to any mRNA sequence, the results are easier to interpret, and it is more rapid (Leban et al., 2024).

10.2.4.5 dsRNA

dsRNA is an unwanted byproduct formed during IVT. These dsRNA species closely resemble viral RNA structures and are recognized by innate immune receptors such as TLR3, MDA5 and RIG-I, triggering inflammatory responses that can reduce mRNA translation efficiency and increase toxicity (Clark et al., 2024). Therefore, removing dsRNA impurities is essential to minimize immunogenicity and improve the overall performance of therapeutic mRNA.

Currently established methods consist of immuno-based detection, such as dot-blotting and enzyme-linked immunosorbent assay (ELISA). These assays rely on antibodies that selectively bind to dsRNA while showing no affinity for single-stranded RNA, including the J2, J5, K1 and K2 monoclonal antibodies, and that can detect sequences that are 40 bp or longer (Schönborn et al., 1991). When using dot-blot, RNA is spotted onto a membrane that is then incubated with a dsRNA-specific monoclonal antibody, typically J2, and the dsRNA signal is detected via chemiluminescence (Baiersdörfer et al., 2019; Karikó et al., 2011; Piao et al., 2022). This assay is semi-quantitative, as the signal must be compared to a calibration series of dots generated from known amounts of dsRNA standard. Sandwich ELISA, on the other hand, detects dsRNA using two different antibodies, mainly J2 and K2, although new optimal antibodies were recently discovered (M2 and M5), resulting in a sensitive quantitative detection method (Holland et al., 2024; Liu et al., 2024).

Specific electrophoresis methods, on the other hand, were shown to offer shorter run times and do not require antibodies. For example, a microfluidic electrophoretic device with dual staining was developed to identify and relatively quantify dsRNA contaminants in mRNA vaccine formulations via fluorescence (Peña et al., 2023). Another option is to enzymatically digest single-stranded mRNA before loading into the microfluidic electrophoretic device, leaving only dsRNA fragments intact for quantification (Peña et al., 2024).

Alternatively, chromatography methods allow for lower limits of detection and quantification, as well as a quicker process that can be applied at multiple steps throughout the manufacturing process (Sousa Rosa et al., 2024). One possibility is using IP-RP-HPLC, where either ssRNA is enzymatically digested before quantifying remaining dsRNA (Sousa Rosa et al., 2024) or a sample is directly loaded and the retention time corresponding to dsRNA is quantified (Nwokeoji et al., 2017).

10.2.4.6 Endotoxins

Endotoxin testing is essential for mRNA vaccines because endotoxins are potent inflammatory agents that can trigger strong immune reactions in patients (United States Pharmacopeia, 2012). According to the USP chapter 85 "Bacterial Endotoxins Test", levels can be measured by three techniques: the gel-clot method, which detects endotoxin through gel formation; the turbidimetric method, which measures the increase in turbidity following cleavage of an endogenous substrate; and the chromogenic method, which monitors color development after cleavage of a synthetic peptide–chromogen complex (United States Pharmacopeia, 2012).

10.2.5 Capping Efficiency

Several mRNA elements are essential for stability and translational efficiency, including the 5′ cap, 5′ and 3′ UTRs and the 3′ poly(A) tail. The 5′ cap and poly(A) tail are required for translation initiation, where interactions among the cap, poly(A)-binding protein and eukaryotic initiation factors drive protein synthesis (Guimaraes et al., 2024). More specifically, the cap consists of an N7-methylated guanosine linked to the first nucleotide through a 5′–5′ triphosphate bridge and includes 2′-O-methylation of the first nucleotide, and it enables recognition by the eIF4E initiation complex, supports ribosome recruitment and protects the 5′ end from exonucleolytic degradation (Gilar et al., 2023). Therefore, it is important to quantify capped versus uncapped mRNA species in therapeutic products to ensure their stability and functional efficacy.

The most used analytical approach for measuring 5′ capping efficiency of mRNA is liquid chromatography–mass spectrometry (LC–MS). Across reported methods, the common principle is to generate RNA fragments, either random ones or specifically ones that contains the 5′ end of the mRNA, that are then resolved by LC–MS analysis for quantitative differentiation between capped and uncapped RNA (Webb et al., 2025).

The use of RNase H for cleavage is recommended in draft USP guidelines (United States Pharmacopeia, 2024), and multiple methods are based on this type of cleavage (Beverly et al., 2016; Nguyen et al., 2021). It is often used in combination with biotinylated DNA probes complementary to the 5′ end of the mRNA, allowing for targeted cleavage and enrichment of fragments by streptavidin-based affinity purification (Beverly et al., 2016). A variation of this approach consists in the use of human RNase 4, which reduces cleavage heterogeneity, followed by a similar workflow (Wolf et al., 2022). Different endonucleases can also be used, such as RNase T1 (Gau et al., 2023; Nakayama et al., 2023), nuclease P1 (Galloway et al., 2020; Muthmann et al., 2022), both also recommended by USP guidelines (United States Pharmacopeia, 2024), and RapiZyme MC1 (Addepalli et al., 2025). These do not require knowledge of the sequence; however, they can result in non-specific or off-target cleavage products (Webb et al., 2025).

Strategies that rely on selective cleavage of the 5′ end prior to LC–MS include ribozyme-based approaches, notably hammerhead (Vlatkovic et al., 2022) or pistol ribozymes (Cui et al., 2025). These have a high specificity, meaning they will cleave at a well-defined location in the sequence thus minimizing non-specific fragments (Webb et al., 2025).

A completely different approach is based on selective enzymatic removal of non-capped RNA species. RNA 5′ polyphosphatase is first used to convert non-capped RNAs into 5′-monophosphorylated forms, while capped RNAs are resistant to this reaction (Chiron & Jais, 2017). These monophosphorylated RNAs are then selectively degraded by a terminator exonuclease, leaving only capped mRNA intact (Chiron & Jais, 2017). Remaining RNA and RNA from an untreated control are quantified using multimodal chromatography, and 5′ capping efficiency is calculated from the obtained ratio of capped RNA to total RNA (Pregeljc et al., 2023).

An additional approach is ligation-mediated RT-qPCR. In this method, RNA 5′ polyphosphatase is first used to convert the 5′ triphosphate of uncapped mRNA into a 5′ monophosphate, while capped RNA remains unchanged (Chiron & Jais, 2017). An anchor RNA is then ligated to these 5′-monophosphorylated ends using a DNA splint and an RNA ligase, such that only uncapped mRNA molecules receive the anchor (Ho, 2017). The ligated RNA is reverse-transcribed into cDNA and quantified by qPCR using two primer sets: one targeting the anchor-containing region and another amplifying a region independent of the anchor (Blewett et al., 2011). Capping efficiency is determined from the ratio of anchor-specific amplicons to total mRNA amplicons.

10.2.6 Poly(A) Tail Length

Similar to the cap, the poly(A) tail, an adenine-rich sequence located at the 3′ end of the mRNA, is critical for mRNA stability and translational efficiency (Camperi et al., 2025). In therapeutic or vaccine applications, poly(A) tails can be introduced by encoding the sequence in the plasmid template, incorporating it through PCR-based template generation or adding it post-IVT with a polyadenylase (Di Grandi et al., 2023). Because poly(A) tail heterogeneity affects both clinical performance and product quality, current USP guidance and other literature classify the poly(A) tail as a CQA (European Medicines Agency, 2025). Two strategies are generally used to characterize this CQA: quantifying the proportion of transcripts that contain a poly(A) tail, or more detailed profiling that measures both the percentage and tail length with resolution sufficient to distinguish single-nucleotide differences (Di Grandi et al., 2023).

The length of the poly(A) tail is typically evaluated using LC paired with UV or mass spectrometry (MS) detection. Because these techniques allow for detection of shorter fragments, the 3′ end of the mRNA must first be digested or cleaved. Various approaches have been explored for this purpose. The RNase H method cleaves at a complementary biotin-labeled oligonucleotide hybridized to the 5′-end, but non-specific cleavage and complex steps limit its practicality (Nilsen, 2015). Ribozymes have been developed to cleave at specific 5′-end positions, yet their susceptibility to degradation affects their stability (Wang et al., 2025). RNase T1 selectively cleaves at guanosine residues, thus releasing the poly(A) tail as an isolated fragment while breaking the rest of the mRNA into shorter pieces; however, it generates fragments of similar size to the tail and is not suitable for poly(A) tails containing interruptions (Gilar et al., 2023). More recently, DNAzymes have been used for site-specific cleavage when designed complementary to the target sequence, offering a potentially more stable and precise approach (Wang et al., 2025).

These fragments can then be separated by SEC, resulting in an estimate of the average poly(A)-tail length by comparing the retention time of the fragment to calibration standards composed of oligo-A sequences of known lengths. Separation by ion-pair reversed-phase liquid chromatography (IP-RP LC) is also possible, where multiple poly(A) species that differ by only a few nucleotides are detected (Gilar et al., 2023). Since this approach generates a chromatographic profile that reflects the heterogeneity of poly(A)-tail lengths, suggesting possible nuclease activity, incomplete tailing or instability during manufacturing or storage, it is preferred for quality control, but it is more complex and less robust than SEC (Gilar et al., 2023). Another possibility is to proceed with MS following cleavage (Beverly et al., 2018b). Although it requires a more expensive high-resolution instrument and specialized deconvolution software, it provides direct molecular-weight measurements of poly(A) tail species and precise mass differences between them, so that tail lengths can be detected with single-nucleotide resolution (Gilar et al., 2023). Additionally, CGE, when also using poly(A) length markers, can provide a distribution of cleaved 3′-end fragments (Di Grandi et al., 2023).

Moreover, a dual approach for poly(A) tail length and capping efficiency was recently developed: mRNA was digested at both ends using DNAzymes, allowing for simultaneous capping and tail analysis, thus reducing sample preparation time and making the overall process more efficient (Wang et al., 2025). Another approach explored the simultaneous analysis as well, but in this case nuclease P1 was used to digest into fragments before subsequent LC/MS detection (Menneteau et al., 2025).

On another note, it was demonstrated that ddPCR can reliably quantify the presence of poly(A) sequences across fractionated mRNA but keeping in mind that it provides no information on tail length or heterogeneity (Patel et al., 2023a). As a result, ddPCR is useful for detecting the presence or loss of poly(A) tails, but it cannot reveal degradation patterns, incomplete transcription or other structural features of the tail (Chang et al., 2014).

Up until now, most methods for assessing mRNA poly(A) tails require sample pre-treatment. A recent approach has been developed that incorporates an immobilized ribonuclease (RNase T1) directly before a hydrophilic interaction liquid chromatography (HILIC) column. When intact mRNA is injected, the ribonuclease cleaves it automatically, and the resulting poly(A) fragments are delivered directly into the HILIC column for separation and analysis – all in a continuous workflow (Goyon et al., 2025).

Advances in sequencing technologies have enabled the measurement of poly(A) tail lengths using platforms such as Illumina (Chang et al., 2014), PacBio (Chang et al., 2014) and Nanopore (Garalde et al., 2018). While these sequencing approaches offer high resolution and can detect subtle sequence variations in mRNA without the need for cleavage, they require multiple preparatory steps, including reverse transcription and the addition of adapters, specialized bioinformatics tools to analyze the data and they can introduce potential errors or biases (Di Grandi et al., 2023).

10.2.7 RNA Encapsulation Efficiency

Following encapsulation of mRNA in LNPs, several CQAs must be measured to ensure proper formation of particles and the stability of the particles' properties throughout manufacturing and storage. The main CQAs related to LNPs are encapsulation efficiency, size, polydispersity index (PDI) and potency (European Medicines Agency, 2025).

Encapsulation efficiency is calculated as the ratio of encapsulated mRNA compared to the total mRNA present inside and outside the LNP. The most widespread approach is to measure RNA concentration of samples with and without a surfactant pre-treatment, which ruptures the particle to measure total mRNA. The concentration of RNA in each sample is commonly measured by a fluorescent dye assay (RiboGreen) (McKenzie et al., 2023), which requires less specialized expertise or equipment, but can also be measured by HPLC (Hara et al., 2024), which allows for higher throughput analysis and eliminates the issue of salts and other components interfering with the RiboGreen fluorescence read-out. Additionally, RNase can be used to degrade the unencapsulated RNA before proceeding to detection through the previously mentioned methods or RT-qPCR (Nag et al., 2022).

Alternatively, sample pre-treatment can be avoided by conducting LNP lysis during the analytical process itself in a two-dimensional chromatography approach (Pavlin et al., 2025). In this case, the first column is used to separate LNPs from free mRNA in weakly hydrophobic conditions, while the second column is used to quantify first the free mRNA and then the subsequently eluted encapsulated mRNA with a UV detector. The strongly hydrophobic chemistry of the column and solvents used for the second step dissolve LNPs inline, without any previous sample manipulation required.

Although non-destructive methods to evaluate encapsulated mRNA are less common, UV–Vis slope spectrometry has been shown to accurately measure encapsulated mRNA concentration and differentiate between empty or loaded LNPs without disruption of the particles (Galchenko & Böttger, 2025). Additionally, no reference standards or dilution curves are required. This may be a practical way to confirm quantification of encapsulated RNA in final products and detect leakage thereafter, but it does not provide a precise ratio between encapsulated and total mRNA to evaluate the efficiency of the encapsulation step. Another work similarly used a UV detector to measure encapsulated mRNA without LNP disruption, but incorporation of a field-flow fractionator enabled separate detection of free mRNA within the same analysis to calculate encapsulation efficiency (Gao et al., 2024).

10.2.8 LNP Size, Polydispersity and Surface Charge

LNP size is crucial for the product's pharmacokinetics and is tunable depending on process parameters like encapsulation flow rate (Di et al., 2022; Hong et al., 2024), often ranging from 50 to 200 nm (Pavlin et al., 2025). Most methods that measure size also characterize the PDI, which reports how narrow or broad the size distribution is. Ideally, LNPs should be as monodisperse as possible, with a neutral surface charge and a consistent size during storage (Lüdtke et al., 2025). Improper or prolonged storage conditions may lead to increased size and polydispersity, indicating particle aggregation (Kamiya et al., 2022).

The most widely used method is dynamic light scattering (DLS), which measures the light scattered by particles during Brownian motion to calculate their hydrodynamic radius (McKenzie et al., 2023). This technique conveniently measures size, PDI and zeta potential (surface charge) but may be inaccurate for polydisperse samples as it tends to overestimate populations with a larger radius (Parot et al., 2024). Spatially resolved DLS (SR-DLS) can be used to measure particle size in a continuous flow for implementation in real-time inline monitoring (Nourafkan et al., 2025).

On the other hand, nanoparticle tracking (NTA) and interferometry light microscopy (ILM or VideoDrop) track individual particles through Brownian motion, providing more accurate identification of separate populations of the particles' hydrodynamic radius (Duffrène et al., 2025; Parot et al., 2024; Roesch et al., 2022). These techniques can also report on number or concentration of individual particles in the solution, which is not typically assessed for mRNA–LNPs but may become important in the future. However, NTA may have high variability if the optimal dilution of samples is not properly determined (Parot et al., 2024). Nano-flow cytometry (NanoFCM), on the other hand, is an orthogonal method that also enables single-particle analysis by using intercalating dyes and multiple laser detectors to measure particle size, number/concentration of particles and differentiate between empty and loaded LNPs (Nourafkan et al., 2025).

Multi-angle light scattering (MALS) is a static detector that measures light scattering from several angles, reporting the particle size as radius of gyration rather than hydrodynamic radius (Pavlin et al., 2025). MALS detectors can be coupled to fractionation methods such as SEC or field-flow fractionation to characterize each subpopulation of particles (Gao et al., 2024; Jia et al., 2021). These detectors can also be connected to other LC approaches to provide an all-in-one assay for particle size, PDI, encapsulation efficiency and RNA integrity (Pavlin et al., 2025).

Electron microscopy imaging such as cryo-EM and transmission electron microscopy (TEM) are high-resolution methods that can provide insight characterization of LNP morphology and size (Di et al., 2022; Jia et al., 2021). Nonetheless, these expensive techniques are low-throughput and therefore mostly used during development of new products and processes for investigation and validation (Roesch et al., 2022).

10.2.9 Lipid Identity and Impurities

Aside from the mRNA molecule, it is also important to characterize the lipids present in the mRNA–LNP product to monitor degradation, impurities and the ratio between the four lipids used. Although the lipids cannot be detected by UV detectors (Parot et al., 2024), RP-HPLC can be used in conjunction with charged aerosol detectors (CAD) or evaporative light scattering detectors (ELSD) to quantify each individual lipid and the ratio between them (Mousli et al., 2022). As CAD and ELSD depend on nebulization and evaporation of the sample, they are destructive detection methods and should, therefore, be placed downstream of any other detectors used in the same system. Alternatively, LC–MS can also be implemented (Fekete et al., 2023; Parot et al., 2024).

Identifying the presence of mRNA-lipid adducts is also crucial as these impurities can lead to untranslatable mRNA that lowers target protein expression. Due to the presence of nucleic acids, UV detectors can be used in this case with an RP-HPLC method or with LC/MS (Packer et al., 2021).

10.2.10 Potency

Potency of mRNA vaccines is assessed using cell-based assays that measure expression of the encoded protein antigen following transfection of eukaryotic cells (Sanyal, 2022). In these assays, mammalian cells (e.g., HEK-293 or HeLa) are transfected with the mRNA construct, and antigen expression is quantitatively detected using immunostaining analytical techniques such as flow cytometry or immunoblotting with antigen-specific monoclonal antibodies (Cao et al., 2025; Feller et al., 2025; Li et al., 2023; Patel et al., 2023b; Sanyal, 2022). Since mRNA drug products are formulated into LNPs, the finished formulation can be directly evaluated in cell-based potency assays. Detection of the expressed antigen by monoclonal antibodies provides evidence not only of protein expression but also of preservation of conformationally relevant epitopes, confirming that a functionally active antigen is produced (Sanyal, 2022).

10.3 REAL-TIME MONITORING IN mRNA–LNP MANUFACTURING

Implementation of real-time monitoring methods is essential to improve efficiency and quality of manufacturing processes. Identifying issues through monitoring can lead to corrective action or early batch termination to avoid losing more time and material. This is especially critical in pandemic-response situations, in which a manufacturing delay of even a few days may largely impact immunization efforts (Schmitt, 2025). In combination with predictive models, these technologies enable digital twin approaches to monitor quality attributes and alter process parameters in a closed feedback loop (Shahab et al., 2025), which is particularly useful for more complex manufacturing approaches such as continuous mode (Schmidt et al., 2022). Importantly, real-time release testing can be achieved through direct real-time measurements of CQAs or prediction of CQAs through models using other real-time measurements (Jiang et al., 2017).

In the context of mRNA manufacturing, several real-time monitoring methods have been investigated and/or are currently used in industry (Table 10.2). Aside from well-established sensors to monitor physicochemical process parameters such as pH, temperature and pressure in multiple steps of the process, advancements have been made to monitor more complex biological quality attributes such as concentration of reagents, concentration of the mRNA product and characterization of mRNA–LNPs. Spectroscopy methods are particularly promising for real-time monitoring of biological molecules as they are non-destructive and can provide information on a variety of attributes through modeling.

IVT can be a complex step to monitor, as the mix of several different components (nucleotides, enzymes, salts, RNA byproducts, RNA product) implicate a high background for simple approaches such as UV absorbance for RNA concentration. However, UV–Vis detectors have been used in conjunction with MALS to detect mRNA concentration in a plug-flow reactor set-up for continuous IVT (Schmidt et al., 2022). Other techniques such as flow-NMR (Sarkar et al., 2024) and Raman spectroscopy (Clenet & Potisopon, 2024; Schmitt, 2025) have been implemented to characterize not only mRNA concentration but also concentration of nucleotides and cap analogs throughout the reaction, enabling optimization of reaction parameters and fed-batch approaches. Raman signals have also been correlated with the concentration of other reagents such as RNAP and different buffer salts, allowing for confirmation of correct concentrations and immediate corrective action if needed (Clenet & Potisopon, 2024; Schmitt, 2025). Additionally, offline Raman measurements have been used to identify point mutations (Morla-Folch et al., 2016),

TABLE 10.2
Established and Forthcoming Methods for Real-Time Monitoring Throughout the mRNA Manufacturing Process

Process Step	Method or Instrument	Analytes or Process Parameters Measured	Reference
***In vitro* transcription**	Thermometer	Temperature	NA
	pHmeter	pH	NA
		Reaction kinetics	Stover (2025)
	Flow-NMR	ATP, CTP, GTP, UTP concentration Cap analog concentration mRNA and phosphate concentration	Sarkar (2024)
	Raman spectroscopy	ATP, CTP, GTP, UTP concentration Cap analog concentration mRNA and phosphate concentration RNA polymerase and glycerol concentration Buffer components (tris, acetate)	Clenet (2024) Schmitt (2025)
	IR and turbidimetry	MgPPi precipitates	Clenet (2024)
	MALS	mRNA concentration	Schmidt (2022)
	UV–Vis	mRNA concentration	Schmidt (2022)
Chromatography purification	Pressure sensor	Pressure	NA
	pHmeter	pH	NA
	Conductivity meter	Conductivity	NA
	UV absorbance	Concentration of eluted products (impurities, mRNA)	NA
Filtration	Pressure sensor	Pressure	NA
	UV–Vis	mRNA concentration	Schmidt (2022)
Encapsulation	SR-DLS	LNP size and PDI	Nourafkan (2025)
	Raman, FTIR, UV–Vis, fluorescence	LNP concentration, size and encapsulation efficiency	Schmidt (2022)

NMR: nuclear magnetic resonance; **IR:** infrared; **MALS:** multi-angle light scattering; **UV–Vis:** ultraviolet-visible; **UV:** ultraviolet; **SR-DLS:** spatially resolves dynamic light scattering; **FTIR:** Fourier-transformed infrared; **MgPPi:** magnesium pyrophosphate; **LNP:** lipid nanoparticle; **PDI:** polydispersity index.

poly(A) tail length (Khadka, 2025) and dsRNA (Morla-Folch et al., 2016) in synthetic oligo mixtures, demonstrating the potential of this technology to become an all-in-one sensor for process parameters, product yield and quality. The formation of MgPPi precipitates which inhibit the reaction have also been tracked using a combination of turbidimetry, infra-red (IR) and Raman spectroscopy (Clenet & Potisopon, 2024). Due to the acidification of the reaction each time a nucleotide is incorporated into an mRNA molecule, pH sensors can also be used to track reaction kinetics (Stover et al., 2025).

LC systems used for purification steps commonly have sensors for pressure, temperature and conductivity to ensure correct operating parameters and buffer characteristics, as well as UV absorbance detectors to measure proteins, nucleotides, DNA and RNA that exit the column at each step of the separation. This can be used to monitor the impurities in the flow-through and wash steps as well as the eluted product.

Filtration is used in several steps of the manufacturing process for product concentration, buffer exchange and removal of impurities and microorganisms. Pressure sensors are typically built in to ensure adequate operating parameters, but UV–Vis can also be implemented to monitor mRNA concentration (Schmidt et al., 2022).

Last, inline or online detectors can be placed downstream of encapsulation and/or of the subsequent filtration to characterize the mRNA–LNP produced in real-time. This was successfully achieved with SR-DLS (Nourafkan et al., 2025) and a multi-PAT detector was proposed using Raman, FTIR, UV–Vis and fluorescence, although this is still under development (Schmidt et al., 2022).

10.4 CONCLUSION AND FUTURE TRENDS

In summary, advanced PAT is important to support efficient and fast production of high-quality mRNA products, as is the case with other biotherapeutics. New modalities of mRNA–LNP therapeutics such as longer self-amplifying RNA (saRNA) molecules, encapsulation with non-lipidic carriers and multivalent vaccines or other drugs with more than one mRNA sequence may require further analytical development to address specific challenges.

Although guidelines have already been drafted with recommendations for quality control in mRNA manufacturing, new methods are still being developed and improved to allow for more efficient, high-resolution and convenient analyses. Certain techniques with the potential of measuring multiple quality attributes in a single assay are particularly interesting to reduce analytical time and labor, including the following: LC for encapsulation efficiency, mRNA integrity, LNP size and PDI (Pavlin et al., 2025); Raman spectroscopy for reagent concentrations (Clenet & Potisopon, 2024; Schmitt, 2025), mRNA concentration (Clenet & Potisopon, 2024; Schmitt, 2025), sequence identity (Morla-Folch et al., 2016), poly(A) tail length (Khadka, 2025) and dsRNA content (Morla-Folch et al., 2016); nanopore sequencing for sequence identity, poly(A) tail length, mRNA integrity and purity (Gunter et al., 2023).

Further development of real-time monitoring methods, as described in the section "Real-time monitoring in mRNA-LNP manufacturing", is also crucial to support the shift to digital twin systems (Helgers et al., 2021; Hengelbrock et al., 2024; Shahab et al., 2025) and integrated continuous manufacturing for mRNA–LNP production (Rosa et al., 2021; Schmidt et al., 2022). This will ultimately lead to shorter and automated processing, increased productivity and higher product quality to further improve the mRNA therapeutics platform.

REFERENCES

Addepalli, B., Johnston, T., Doneanu, C., Vaishnav, J., Gomes, A., Reidy, C., Yu, Y. Q., & Lauber, M. A. (2025). *Analysis of mRNA Cap Impurity Profiles and Capping Efficiency Using RapiZyme™ MC1 Ribonuclease*. Application note Waters. https://www.waters.com/content/dam/waters/en/app-notes/2025/720008793/720008793-en.pdf

Agbogbo, F., George, R., Krishnamoorthy, G., Davis, C., & Farrand, A. (2023). Process Development for Plasmid DNA Production. In *Process Validation in Manufacturing of Biopharmaceuticals* (4th ed.). CRC Press.

Ahl, P. L. (2025). Microfluidic and turbulent mixing for mRNA LNP vaccines. *Pharmaceutics*, *17*(9). https://doi.org/10.3390/pharmaceutics17091148

American Society of Gene and Cell Therapy. (2025, August 1). *Q2 2025 Gene, Cell + RNA Therapy Landscape Report.* Citeline. https://www.citeline.com/en/resources/q2-2025-gene-cell-and-rna-therapy-report

Baiersdörfer, M., Boros, G., Muramatsu, H., Mahiny, A., Vlatkovic, I., Sahin, U., & Karikó, K. (2019). A facile method for the removal of dsRNA contaminant from in vitro-transcribed mRNA. *Molecular Therapy. Nucleic Acids*, *15*, 26–35. https://doi.org/10.1016/j.omtn.2019.02.018

Barros, C. H. N., Alfaro, M., Csiki-Fejer, A., Bhatnagar, B., Tschessalov, S., Ferguson, S., Barua, S., Darvari, R., & Topp, E. M. (2025). Comparative analysis of mRNA degradation kinetics using chromatographic and electrophoretic methods. *Molecular Pharmaceutics*, *22*(6), 3061–3072. https://doi.org/10.1021/acs.molpharmaceut.4c01543

Beverly, M., Dell, A., Parmar, P., & Houghton, L. (2016). Label-free analysis of mRNA capping efficiency using RNase H probes and LC-MS. *Analytical and Bioanalytical Chemistry*, *408*(18), 5021–5030. https://doi.org/10.1007/s00216-016-9605-x

Beverly, M., Hagen, C., & Slack, O. (2018a). Poly A tail length analysis of in vitro transcribed mRNA by LC-MS. *Analytical and Bioanalytical Chemistry*, *410*(6), 1667–1677. https://doi.org/10.1007/s00216-017-0840-6

Beverly, M., Hagen, C., & Slack, O. (2018b). Poly A tail length analysis of in vitro transcribed mRNA by LC-MS. *Analytical and Bioanalytical Chemistry*, *410*(6), 1667–1677. https://doi.org/10.1007/s00216-017-0840-6

Blewett, N., Coller, J., & Goldstrohm, A. (2011). A quantitative assay for measuring mRNA decapping by splinted ligation reverse transcription polymerase chain reaction: qSL-RT-PCR. *RNA*, *17*(3), 535–543. https://doi.org/10.1261/rna.2436411

Camperi, J., Chatla, K., Freund, E., Galan, C., Lippold, S., & Guilbaud, A. (2025). Current analytical strategies for mRNA-based therapeutics. *Molecules (Basel, Switzerland)*, *30*(7), 1629. https://doi.org/10.3390/molecules30071629

Camperi, J., Lippold, S., Ayalew, L., Roper, B., Shao, S., Freund, E., Nissenbaum, A., Galan, C., Cao, Q., Yang, F., Yu, C., & Guilbaud, A. (2024). Comprehensive impurity profiling of mRNA: Evaluating current technologies and advanced analytical techniques. *Analytical Chemistry*, *96*(9), 3886–3897. https://doi.org/10.1021/acs.analchem.3c05539

Cao, F., Xu, Y., Guan, Y., Zhang, K., Qiu, H., Xu, Z., He, Y., Xiao, Z. X., Zha, G., & Pang, J. (2025). Enhancing the potency of 5T4 mRNA vaccine by CD70 mRNA-LNPs through ADCC and T cell boosting in prostate cancer therapy. *Journal of Nanobiotechnology*, *23*(1), 523. https://doi.org/10.1186/s12951-025-03607-4

Chang, H., Lim, J., Ha, M., & Kim, V. N. (2014). TAIL-seq: Genome-wide determination of poly(A) tail length and 3′ end modifications. *Molecular Cell*, *53*(6), 1044–1052. https://doi.org/10.1016/j.molcel.2014.02.007

Chiron, S., & Jais, P. (2017). *Non-Radioactive Monitoring Assay for Capping of Messenger RNA. Recherche Google.* https://www.google.com/search?client=safari&rls=en&q=Non-radioactive+monitoring+assay+for+capping+of+messenger+RNA&ie=UTF-8&oe=UTF-8

Clark, N. E., Schraut, M. R., Winters, R. A., Kearns, K., & Scanlon, T. C. (2024). An immunonorthern technique to measure the size of dsRNA byproducts in in vitro transcribed RNA. *Electrophoresis*, *45*(17–18), 1546–1554. https://doi.org/10.1002/elps.202400036

Clenet, D., & Potisopon, S. (2024). *Spectral Monitoring of In Vitro Transcription* (World Intellectual Property Organization Patent No. WO2024074726A1). https://patents.google.com/patent/WO2024074726A1/en

Cui, T., Fakhfakh, K., Turney, H., Güler-Gane, G., Toloczko, A., Hulley, M., & Turner, R. (2023). Comprehensive studies on building a scalable downstream process for mRNAs to enable mRNA therapeutics. *Biotechnology Progress*, *39*(1), e3301. https://doi.org/10.1002/btpr.3301

Cui, Y., Chen, T., Zhang, Y., Zhang, G., & Lu, C. (2025). Capping in mRNA analyzed using pistol ribozyme and LC-MS. *Biochemical and Biophysical Research Communications*, *766*, 151860. https://doi.org/10.1016/j.bbrc.2025.151860

Daniel, S., Kis, Z., Kontoravdi, C., & Shah, N. (2022). Quality by design for enabling RNA platform production processes. *Trends in Biotechnology*, *40*(10), 1213–1228. https://doi.org/10.1016/j.tibtech.2022.03.012

Di Grandi, D., Dayeh, D. M., Kaur, K., Chen, Y., Henderson, S., Moon, Y., Bhowmick, A., Ihnat, P. M., Fu, Y., Muthusamy, K., Palackal, N., & Pyles, E. A. (2023). A single-nucleotide resolution capillary gel electrophoresis workflow for poly(A) tail characterization in the development of mRNA therapeutics and vaccines. *Journal of Pharmaceutical and Biomedical Analysis*, *236*, 115692. https://doi.org/10.1016/j.jpba.2023.115692

Di, J., Du, Z., Wu, K., Jin, S., Wang, X., Li, T., & Xu, Y. (2022). Biodistribution and nonlinear gene expression of mRNA LNPs affected by delivery route and particle size. *Pharmaceutical Research*, *39*(1), 105–114. https://doi.org/10.1007/s11095-022-03166-5

Duffrène, J., Muzard, C., Seguin, J., Izabelle, C., Vrai, T., Ejlalmanesh, T., Bombled, M., Hamdi, S., Lemdani, K., Alhareth, K., & Mignet, N. (2025). Post-encapsulation methods for the preparation of mRNA-LNPs. *Drug Delivery and Translational Research*, *15*(12), 4729–4741. https://doi.org/10.1007/s13346-025-01866-0

European Medicines Agency. (2025). *Guideline on the Quality Aspects of mRNA Vaccines* [Guidelines]. Committee for Medicinal Products for Human Use (CHMP).

Fekete, S., Doneanu, C., Addepalli, B., Gaye, M., Nguyen, J., Alden, B., Birdsall, R., Han, D., Isaac, G., & Lauber, M. (2023). Challenges and emerging trends in liquid chromatography-based analyses of mRNA pharmaceuticals. *Journal of Pharmaceutical and Biomedical Analysis*, *224*, 115174. https://doi.org/10.1016/j.jpba.2022.115174

Feller, K., Nawar, H., Song, L., Abrams, A., Shang, L., Gruber, A., Yun, T., Li, H. H., Feller, K., Nawar, H., Song, L., Abrams, A., Shang, L., Gruber, A., Yun, T., & Li, H. H. (2025). Cell-based relative potency of a respiratory syncytial virus mRNA vaccine correlates with in vivo immunogenicity. *Vaccines*, *13*(3). https://doi.org/10.3390/vaccines13030326

Food and Drug Administration. (2021, May 26). *Direct Analysis of Residual Host Cell DNA by Droplet Digital PCR (ddPCR) in Biologic Drugs Produced in E. Coli*. FDA. https://www.fda.gov/science-research/fda-science-forum/direct-analysis-residual-host-cell-dna-droplet-digital-pcr-ddpcr-biologic-drugs-produced-e-coli

Galchenko, M., & Böttger, R. (2025). *Rapid and Nondestructive Total mRNA Quantification with Slope Spectroscopy* ***for Conventional and UV-Absorbing Lipid Nanoparticles***. Bioprocess International 23(3). https://eu-assets.contentstack.com/v3/assets/blt0a48a1f3edca9eb0/bltfad1d2856c58fb83/67a64c356f83485c3d86d261/23-3-CR-Repligen.pdf

Galloway, A., Atrih, A., Grzela, R., Darzynkiewicz, E., Ferguson, M. A. J., & Cowling, V. H. (2020). CAP-MAP: Cap analysis protocol with minimal analyte processing, a rapid and sensitive approach to analysing mRNA cap structures. *Open Biology*, *10*(2), 190306. https://doi.org/10.1098/rsob.190306

Gao, Z., Lin, J., Su, W.-C., Zhang, K., Gruenhagen, J., Zhong, W., Fan, Y., & Bian, J. (2024). Development of an advanced separation and characterization platform for mRNA and lipid nanoparticles using multi-detector asymmetrical flow field-flow fractionation. *Analytical and Bioanalytical Chemistry*, *416*(24), 5281–5293. https://doi.org/10.1007/s00216-024-05455-x

Garalde, D. R., Snell, E. A., Jachimowicz, D., Sipos, B., Lloyd, J. H., Bruce, M., Pantic, N., Admassu, T., James, P., Warland, A., Jordan, M., Ciccone, J., Serra, S., Keenan,

J., Martin, S., McNeill, L., Wallace, E. J., Jayasinghe, L., Wright, C., … Turner, D. J. (2018). Highly parallel direct RNA sequencing on an array of nanopores. *Nature Methods*, *15*(3), 201–206. https://doi.org/10.1038/nmeth.4577

Gau, B. C., Dawdy, A. W., Wang, H. L., Bare, B., Castaneda, C. H., Friese, O. V., Thompson, M. S., Lerch, T. F., Cirelli, D. J., & Rouse, J. C. (2023). Oligonucleotide mapping via mass spectrometry to enable comprehensive primary structure characterization of an mRNA vaccine against SARS-CoV-2. *Scientific Reports*, *13*(1), 9038. https://doi.org/10.1038/s41598-023-36193-2

Gilar, M., Doneanu, C., & Gaye, M. M. (2023). Liquid chromatography methods for analysis of mRNA poly(A) tail length and heterogeneity. *Analytical Chemistry*, *95*(38), 14308–14316. https://doi.org/10.1021/acs.analchem.3c02552

Goyon, A., Nguyen, D. N., Wang, S., Hofmann, K., & Zhang, K. (2025). Online determination of mRNA poly(A) tail length and content. *Analytical Chemistry*, *97*(39), 21572–21579. https://doi.org/10.1021/acs.analchem.5c03982

Green, M. R., & Sambrook, J. (2020). Quantifying and storing RNA. *Cold Spring Harbor Protocols*, *2020*(3), pdb.top101709. https://doi.org/10.1101/pdb.top101709

Guimaraes, G. J., Kim, J., & Bartlett, M. G. (2024). Characterization of mRNA therapeutics. *Mass Spectrometry Reviews*, *43*(5), 1066–1090. https://doi.org/10.1002/mas.21856

Gunter, H. M., Idrisoglu, S., Singh, S., Han, D. J., Ariens, E., Peters, J. R., Wong, T., Cheetham, S. W., Xu, J., Rai, S. K., Feldman, R., Herbert, A., Marcellin, E., Tropee, R., Munro, T., & Mercer, T. R. (2023). mRNA vaccine quality analysis using RNA sequencing. *Nature Communications*, *14*(1), 5663. https://doi.org/10.1038/s41467-023-41354-y

Hara, S., Arase, S., Sano, S., Suzuki, T., Mizogaki, I., Sato, S., & Ukai, K. (2024). Anion exchange-HPLC method for evaluating the encapsulation efficiency of mRNA-loaded lipid nanoparticles using analytical quality by design. *Journal of Chromatography B*, *1247*, 124317. https://doi.org/10.1016/j.jchromb.2024.124317

Helgers, H., Hengelbrock, A., Schmidt, A., Strube, J., Helgers, H., Hengelbrock, A., Schmidt, A., & Strube, J. (2021). Digital twins for continuous mRNA production. *Processes*, *9*(11). https://doi.org/10.3390/pr9111967

Hengelbrock, A., Schmidt, A., & Strube, J. (2024). Digital twin fundamentals of mRNA in vitro transcription in variable scale toward autonomous operation. *ACS Omega*, *9*(7), 8204–8220. https://doi.org/10.1021/acsomega.3c08732

Hersch, S. J., Chandrasekaran, S., Lam, J., Nafissi, N., & Slavcev, R. A. (2024). Manufacturing DNA in *E. coli* yields higher-fidelity DNA than in vitro enzymatic synthesis. *Molecular Therapy Methods & Clinical Development*, *32*(2). https://doi.org/10.1016/j.omtm.2024.101227

Ho, C. K. (2017). Detection and identification of uncapped RNA by ligation-mediated reverse transcription polymerase chain reaction. *Methods in Molecular Biology (Clifton, N.J.)*, *1648*, 1–9. https://doi.org/10.1007/978-1-4939-7204-3_1

Holland, D. A., Acevedo-Skrip, J., Barton, J., Thompson, R., Bowman, A., Dewar, E. A., Miller, D. V., Zhao, K., Swartz, A. R., Loughney, J. W., Holland, D. A., Acevedo-Skrip, J., Barton, J., Thompson, R., Bowman, A., Dewar, E. A., Miller, D. V., Zhao, K., Swartz, A. R., & Loughney, J. W. (2024). Development and application of automated sandwich ELISA for quantitating residual dsRNA in mRNA vaccines. *Vaccines*, *12*(8). https://doi.org/10.3390/vaccines12080899

Hong, J., Lee, S., Park, H., Ahn, D., Lee, J. M., Choe, H., Kim, D., Kim, J. H., & Chon, C. H. (2024). Size-controllable and monodispersed lipid nanoparticle production with high mRNA delivery efficiency using 3D-printed ring micromixers. *ACS Applied Materials & Interfaces*, *16*(35), 46044–46052. https://doi.org/10.1021/acsami.4c08896

Hu, C., Bai, Y., Liu, J., Wang, Y., He, Q., Zhang, X., Cheng, F., Xu, M., Mao, Q., & Liang, Z. (2024). Research progress on the quality control of mRNA vaccines. *Expert Review of Vaccines*, *23*(1), 570–583. https://doi.org/10.1080/14760584.2024.2354251

Jia, X., Liu, Y., Wagner, A. M., Chen, M., Zhao, Y., Smith, K. J., Some, D., Abend, A. M., & Pennington, J. (2021). Enabling online determination of the size-dependent RNA content of lipid nanoparticle-based RNA formulations. *Journal of Chromatography B*, *1186*, 123015. https://doi.org/10.1016/j.jchromb.2021.123015

Jiang, M., Severson, K. A., Love, J. C., Madden, H., Swann, P., Zang, L., & Braatz, R. D. (2017). Opportunities and challenges of real-time release testing in biopharmaceutical manufacturing. *Biotechnology and Bioengineering*, *114*(11), 2445–2456. https://doi.org/10.1002/bit.26383

Jiang, T., Yu, N., Kim, J., Murgo, J.-R., Kissai, M., Ravichandran, K., Miracco, E. J., Presnyak, V., & Hua, S. (2019). Oligonucleotide sequence mapping of large therapeutic mRNAs via parallel ribonuclease digestions and LC-MS/MS. *Analytical Chemistry*, *91*(13), 8500–8506. https://doi.org/10.1021/acs.analchem.9b01664

Kafle, U., Truong, H. Q., Nguyen, C. T. G., & Meng, F. (2024). Development of thermally stable mRNA-LNP delivery systems: Current progress and future prospects. *Molecular Pharmaceutics*, *21*(12), 5944–5959. https://doi.org/10.1021/acs.molpharmaceut.4c00826

Kamiya, M., Matsumoto, M., Yamashita, K., Izumi, T., Kawaguchi, M., Mizukami, S., Tsurumaru, M., Mukai, H., Kawakami, S., Kamiya, M., Matsumoto, M., Yamashita, K., Izumi, T., Kawaguchi, M., Mizukami, S., Tsurumaru, M., Mukai, H., & Kawakami, S. (2022). Stability study of mRNA-lipid nanoparticles exposed to various conditions based on the evaluation between physicochemical properties and their relation with protein expression ability. *Pharmaceutics*, *14*(11). https://doi.org/10.3390/pharmaceutics14112357

Karikó, K., Muramatsu, H., Ludwig, J., & Weissman, D. (2011). Generating the optimal mRNA for therapy: HPLC purification eliminates immune activation and improves translation of nucleoside-modified, protein-encoding mRNA. *Nucleic Acids Research*, *39*(21), e142. https://doi.org/10.1093/nar/gkr695

Khadka, N. (2025, December 15). *Quantification of Adenine Residues in the PolyA Tail of Oligonucleotides Using Raman Spectroscopy. Spectroscopy Online*. https://www.spectroscopyonline.com/view/quantification-of-adenine-residues-in-the-polya-tail-of-oligonucleotides-using-raman-spectroscopy

Korenč, M., Mencin, N., Puc, J., Skok, J., Nemec, K. Š., Celjar, A. M., Gagnon, P., Štrancar, A., & Sekirnik, R. (2021). Chromatographic purification with CIMmultus™ Oligo dT increases mRNA stability. *Cell and Gene Therapy Insights*. https://doi.org/10.18609/cgti.2021.161

Krammer, F. (2024). The role of vaccines in the COVID-19 pandemic: What have we learned? *Seminars in Immunopathology*, *45*(4), 451–468. https://doi.org/10.1007/s00281-023-00996-2

Lauro, M. L., Bowman, A. M., Smith, J. P., Gaye, S. N., Acevedo-Skrip, J., DePhillips, P. A., & Loughney, J. W. (2022). Overcoming biopharmaceutical interferents for quantitation of host cell DNA using an automated, high-throughput methodology. *The AAPS Journal*, *25*(1), 10. https://doi.org/10.1208/s12248-022-00764-4

Leban, M., Vodopivec Seravalli, T., Hauer, M., Böhm, E., Mencin, N., Potušek, S., Thompson, A., Trontelj, J., Štrancar, A., & Sekirnik, R. (2024). Determination of linearized pDNA template in mRNA production process using HPLC. *Analytical and Bioanalytical Chemistry*, *416*(10), 2389–2398. https://doi.org/10.1007/s00216-024-05204-0

Lee, D. H., Bae, J. E., Lee, J. H., Shin, J. S., & Kim, I. S. (2010). Quantitative detection of residual *E. coli* host cell DNA by real-time PCR. *Journal of Microbiology and Biotechnology*, *20*(10), 1463–1470. https://doi.org/10.4014/jmb.1004.04035

Li, D., Zhang, Q., Liu, G., Zhang, L., Gu, Z., Pan, Y., Cui, X., He, P., Li, X., Liu, J., Liu, G., Yang, M., & Tian, X. (2021). Detection of residual *E. coli* host cell DNA by 23S ribosomal RNA gene-targeted quantitative polymerase chain reactions. *Journal of Pharmaceutical and Biomedical Analysis*, *198*, 114000. https://doi.org/10.1016/j.jpba.2021.114000

Li, H. H., Xu, J., He, L., Denny, L. I., Rustandi, R. R., Dornadula, G., Fiorito, B., & Zhang, Z.-Q. (2023). Development and qualification of cell-based relative potency assay for a human respiratory syncytial virus (RSV) mRNA vaccine. *Journal of Pharmaceutical and Biomedical Analysis*, *234*, 115523. https://doi.org/10.1016/j.jpba.2023.115523

Liu, J., Zheng, T., Xu, L., Chen, Z., Zhang, K., Wang, X., Xu, X., Li, Y., Sun, Y., & Zhu, L. (2024). An improved method for the detection of double-stranded RNA suitable for quality control of mRNA vaccines. *Protein & Cell*, *15*(11), 791–795. https://doi.org/10.1093/procel/pwae043

Lüdtke, F. L., Silva, T. J., Silva, M. G. da, Hashimoto, J. C., Ribeiro, A. P. B., Lüdtke, F. L., Silva, T. J., Silva, M. G. da, Hashimoto, J. C., & Ribeiro, A. P. B. (2025). Lipid nanoparticles: Formulation, production methods and characterization protocols. *Foods*, *14*(6). https://doi.org/10.3390/foods14060973

Madigan, J. (2021). *Rapid mRNA Manufacturing for Vaccine Development*. https://www.trilinkbiotech.com/media/maravai/productattachments/coa/PT0921_TriLink_082421_ES_004_.pdf

Mantri, P., Juneja, B., Henderson, S., Koufos, E., Moon, Y., Dayeh, D. M., Di Grandi, D., Fu, Y., Muthusamy, K., Ihnat, P. M., Palackal, N., & Pyles, E. A. (2024). Comparison of capillary electrophoresis-based methods for the analytical characterization of purity and stability of *in vitro* transcribed mRNA. *Journal of Pharmaceutical and Biomedical Analysis*, *249*, 116352. https://doi.org/10.1016/j.jpba.2024.116352

McKenzie, R. E., Minnell, J. J., Ganley, M., Painter, G. F., & Draper, S. L. (2023). mRNA synthesis and encapsulation in ionizable lipid nanoparticles. *Current Protocols*, *3*(9), e898. https://doi.org/10.1002/cpz1.898

Menneteau, T., Butré, C. I., Mouvet, D., & Delobel, A. (2025, December 15). *Multi-Attribute Monitoring of Therapeutic mRNA by Liquid Chromatography–Mass Spectrometry. LCGC International.* https://www.chromatographyonline.com/view/multi-attribute-monitoring-therapeutic-mrna-liquid-chromatography-mass-spectrometry

Morla-Folch, J., Xie, H., Alvarez-Puebla, R. A., & Guerrini, L. (2016). Fast optical chemical and structural classification of RNA. *ACS Nano*, *10*(2), 2834–2842. https://doi.org/10.1021/acsnano.5b07966

Mousli, Y., Brachet, M., Chain, J. L., & Ferey, L. (2022). A rapid and quantitative reversed-phase HPLC-DAD/ELSD method for lipids involved in nanoparticle formulations. *Journal of Pharmaceutical and Biomedical Analysis*, *220*, 115011. https://doi.org/10.1016/j.jpba.2022.115011

Muthmann, N., Špaček, P., Reichert, D., van Dülmen, M., & Rentmeister, A. (2022). Quantification of mRNA cap-modifications by means of LC-QqQ-MS. *Methods*, *203*, 196–206. https://doi.org/10.1016/j.ymeth.2021.05.018

Nag, K., Sarker, M. E. H., Kumar, S., Khan, H., Chakraborty, S., Islam, M. J., Baray, J. C., Khan, M. R., Mahmud, A., Barman, U., Bhuiya, E. H., Mohiuddin, M., & Sultana, N. (2022). DoE-derived continuous and robust process for manufacturing of pharmaceutical-grade wide-range LNPs for RNA-vaccine/drug delivery. *Scientific Reports*, *12*(1), 9394. https://doi.org/10.1038/s41598-022-12100-z

Nakayama, H., Nobe, Y., Koike, M., & Taoka, M. (2023). Liquid chromatography–mass spectrometry-based qualitative profiling of mRNA therapeutic reagents using stable isotope-labeled standards followed by the automatic quantitation software Ariadne. *Analytical Chemistry*, *95*(2), 1366–1375. https://doi.org/10.1021/acs.analchem.2c04323

Nguyen, J. M., Chan, S.-H., Roy, B., Gilar, M., Robb, B., Chen, W., & Lauber, M. A. (2021). *Rapid Analysis of Synthetic mRNA Cap Structure Using Ion-Pairing RPLC with the BioAccord LC-MS System*. https://www.waters.com/nextgen/ca/en/library/application-notes/2021/rapid-analysis-of-synthetic-mrna-cap-structure-using-ion-pairing-rplc-with-the-bioaccord-lc-ms-system.html?srsltid=AfmBOookTsMuFaaQlIR62nmAITCKBGXwS__X_atAxhxjDyT_AZymid-o

Nilsen, T. W. (2015). Measuring the length of poly(A) tails. *Cold Spring Harbor Protocols, 2015*(4), 413–418. https://doi.org/10.1101/pdb.prot081034

Nourafkan, E., Yang, Z., Maamra, M., & Kis, Z. (2025). Advancing continuous encapsulation and purification of mRNA vaccines and therapeutics. *European Journal of Pharmaceutical Sciences, 212*, 107183. https://doi.org/10.1016/j.ejps.2025.107183

Nwokeoji, A. O., Kung, A.-W., Kilby, P. M., Portwood, D. E., & Dickman, M. J. (2017). Purification and characterisation of dsRNA using ion pair reverse phase chromatography and mass spectrometry. *Journal of Chromatography A, 1484*, 14–25. https://doi.org/10.1016/j.chroma.2016.12.062

Packer, M., Gyawali, D., Yerabolu, R., Schariter, J., & White, P. (2021). A novel mechanism for the loss of mRNA activity in lipid nanoparticle delivery systems. *Nature Communications, 12*(1), 6777. https://doi.org/10.1038/s41467-021-26926-0

Pardi, N., & Krammer, F. (2024). mRNA vaccines for infectious diseases—Advances, challenges and opportunities. *Nature Reviews Drug Discovery, 23*(11), 838–861. https://doi.org/10.1038/s41573-024-01042-y

Parhiz, H., Atochina-Vasserman, E. N., & Weissman, D. (2024). mRNA-based therapeutics: Looking beyond COVID-19 vaccines. *The Lancet, 403*(10432), 1192–1204. https://doi.org/10.1016/S0140-6736(23)02444-3

Parot, J., Mehn, D., Jankevics, H., Markova, N., Carboni, M., Olaisen, C., Hoel, A. D., Sigfúsdóttir, M. S., Meier, F., Drexel, R., Vella, G., McDonagh, B., Hansen, T., Bui, H., Klinkenberg, G., Visnes, T., Gioria, S., Urban-Lopez, P., Prina-Mello, A., … Calzolai, L. (2024). Quality assessment of LNP-RNA therapeutics with orthogonal analytical techniques. *Journal of Controlled Release, 367*, 385–401. https://doi.org/10.1016/j.jconrel.2024.01.037

Patel, H. K., Zhang, K., Utegg, R., Stephens, E., Salem, S., Welch, H., Grobe, S., Schlereth, J., Kuhn, A. N., Ryczek, J., Cirelli, D. J., & Lerch, T. F. (2023a). Characterization of BNT162b2 mRNA to evaluate risk of off-target antigen translation. *Journal of Pharmaceutical Sciences, 112*(5), 1364–1371. https://doi.org/10.1016/j.xphs.2023.01.007

Patel, N., Davis, Z., Hofmann, C., Vlasak, J., Loughney, J. W., DePhillips, P., Mukherjee, M., Patel, N., Davis, Z., Hofmann, C., Vlasak, J., Loughney, J. W., DePhillips, P., & Mukherjee, M. (2023b). Development and characterization of an in vitro cell-based assay to predict potency of mRNA–LNP-based vaccines. *Vaccines, 11*(7). https://doi.org/10.3390/vaccines11071224

Pavlin, N., Bavčar, M., Kovačič, T., Kašček, T., Celjar, A. M., Bergoč, I., Livk, A. G., & Štrancar, A. (2025). Analysis of lipid nanoparticles using two-dimensional chromatography: Simultaneous determination of encapsulation efficiency, nucleic acid integrity, and size of LNP formulations. *Journal of Chromatography B, 1265*, 124751. https://doi.org/10.1016/j.jchromb.2025.124751

Peña, A. C. D., Li, N., Vaduva, M., Bwanali, L., & Tripathi, A. (2023). A microfluidic electrophoretic dual dynamic staining method for the identification and relative quantitation of dsRNA contaminants in mRNA vaccines. *Analyst, 148*(16), 3758–3767. https://doi.org/10.1039/D3AN00281K

Peña, A. C. D., Vaduva, M., Li, N. S., Shah, S., Frej, M. B., & Tripathi, A. (2024). Enzymatic isolation and microfluidic electrophoresis analysis of residual dsRNA impurities in mRNA vaccines and therapeutics. *Analyst, 149*(5), 1509–1517. https://doi.org/10.1039/D3AN02157B

Peper, G., Fankhauser, A., Merlin, T., Roscic, A., Hofmann, M., & Obrdlik, P. (2014). Direct real-time quantitative PCR for measurement of host-cell residual DNA in therapeutic proteins. *Journal of Pharmaceutical and Biomedical Analysis, 100*, 123–130. https://doi.org/10.1016/j.jpba.2014.07.032

Piao, X., Yadav, V., Wang, E., Chang, W., Tau, L., Lindenmuth, B. E., & Wang, S. X. (2022). Double-stranded RNA reduction by chaotropic agents during in vitro transcription of messenger RNA. *Molecular Therapy. Nucleic Acids*, *29*, 618–624. https://doi.org/10.1016/j.omtn.2022.08.001

Pregeljc, D., Skok, J., Vodopivec, T., Mencin, N., Krušič, A., Ličen, J., Nemec, K. Š., Štrancar, A., & Sekirnik, R. (2023). Increasing yield of in vitro transcription reaction with at-line high pressure liquid chromatography monitoring. *Biotechnology and Bioengineering*, *120*(3), 737–747. https://doi.org/10.1002/bit.28299

Roesch, A., Zölls, S., Stadler, D., Helbig, C., Wuchner, K., Kersten, G., Hawe, A., Jiskoot, W., & Menzen, T. (2022). Particles in biopharmaceutical formulations, Part 2: An update on analytical techniques and applications for therapeutic proteins, viruses, vaccines and cells. *Journal of Pharmaceutical Sciences*, *111*(4), 933–950. https://doi.org/10.1016/j.xphs.2021.12.011

Rosa, S. S., Prazeres, D. M. F., Azevedo, A. M., & Marques, M. P. C. (2021). mRNA vaccines manufacturing: Challenges and bottlenecks. *Vaccine*, *39*(16), 2190–2200. https://doi.org/10.1016/j.vaccine.2021.03.038

Sanyal, G. (2022). Development of functionally relevant potency assays for monovalent and multivalent vaccines delivered by evolving technologies. *npj Vaccines*, *7*(1), 50. https://doi.org/10.1038/s41541-022-00470-4

Sarkar, A., Dong, G., Quaglia-Motta, J., & Sackett, K. (2024). Flow-NMR as a process-monitoring tool for mRNA IVT reaction. *Journal of Pharmaceutical Sciences*, *113*(4), 900–905. https://doi.org/10.1016/j.xphs.2023.11.021

Schmidt, A., Helgers, H., Vetter, F. L., Juckers, A., Strube, J., Schmidt, A., Helgers, H., Vetter, F. L., Juckers, A., & Strube, J. (2021). Digital twin of mRNA-based SARS-COVID-19 vaccine manufacturing towards autonomous operation for improvements in speed, scale, robustness, flexibility and real-time release testing. *Processes*, *9*(5). https://doi.org/10.3390/pr9050748

Schmidt, A., Helgers, H., Vetter, F. L., Zobel-Roos, S., Hengelbrock, A., & Strube, J. (2022). Process automation and control strategy by quality-by-design in total continuous mRNA manufacturing platforms. *Processes*, *10*(9). https://doi.org/10.3390/pr10091783

Schmitt, E. (2025). *Methods of Monitoring In Vitro Transcription of mRNA and/or Post-In Vitro Transcription Processes* (United States Patent No. US12428577B2). https://patents.google.com/patent/US12428577B2/en

Schönborn, J., Oberstrass, J., Breyel, E., Tittgen, J., Schumacher, J., & Lukacs, N. (1991). Monoclonal antibodies to double-stranded RNA as probes of RNA structure in crude nucleic acid extracts. *Nucleic Acids Research*, *19*(11), 2993–3000. https://doi.org/10.1093/nar/19.11.2993

Shahab, M. A., Stover, N. M., Al-Mahayni, H., Srisuma, P., Wang, J., Shin, S., Dighe, A., Trout, B. L., Myerson, A. S., & Braatz, R. D. (2025). End-to-end digital twin software for continuous mRNA manufacturing. *International Journal of Pharmaceutics*, *686*, 126301. https://doi.org/10.1016/j.ijpharm.2025.126301

Shankar, R., Schmeer, M., & Schleef, M. (2024). Producing Plasmid DNA Template for Clinical Grade RNA Vaccine Manufacture. In T. Kramps (Ed.), *RNA Vaccines: Methods and Protocols* (pp. 303–319). Springer US. https://doi.org/10.1007/978-1-0716-3770-8_14

Shieh, Y., Swartz, A. R., & Rustandi, R. R. (2024). Detection of residual T7 RNA polymerase used in mRNA in vitro transcription by Simple Western. *Electrophoresis*, *45*(19–20), 1834–1839. https://doi.org/10.1002/elps.202400044

Sousa Rosa, S., Zhang, S., Sari, Y., & C. Marques, M. P. (2024). A (RP)UHPLC/UV analytical method to quantify dsRNA during the mRNA vaccine manufacturing process. *Analytical Methods*, *16*(30), 5146–5153. https://doi.org/10.1039/D4AY00560K

Stover, N. M., Ahmadi, S., Rosenfeld, J., Destro, F., Myerson, A. S., & Braatz, R. D. (2025). Model-based optimization of fed-batch in vitro transcription. *ChemBioChem*, *26*(21), e202500485. https://doi.org/10.1002/cbic.202500485

Su, W.-C., Yu, Z., Lieu, R., Freund, E., Hokeness, K., Gu, Y., Kohler, J., Chen, T., Fan, Y., & Zhang, K. (2025). Developing a high-throughput capillary gel electrophoresis workflow for mRNA integrity and poly(A) tail analysis. *Journal of Chromatography A, 1761*, 466387. https://doi.org/10.1016/j.chroma.2025.466387

Tian, Y., Wang, X., Shao, D., Zhao, W., Chen, R., & Huang, Q. (2024). Establishment and evaluation of detection methods for process-specific residual host cell protein and residual host cell DNA in biological preparation. *Cell Biochemistry and Function, 42*(2), e3986. https://doi.org/10.1002/cbf.3986

United States Pharmacopeia. (2012) *Bacterial Endotoxins.* https://www.usp.org/harmonization-standards/pdg/general-methods/bacterial-endotoxins

United States Pharmacopeia. (2024, August 2). *Analytical Procedures for Quality of mRNA Vaccines and Therapeutics.* https://go.usp.org/l/323321/2022-02-08/6rxnz3/323321/16443533303H7od3V5/EA966W_mRNA_Vaccine_Chapter_2022_03_3.pdf

Vanhinsbergh, C. J., Criscuolo, A., Sutton, J. N., Murphy, K., Williamson, A. J. K., Cook, K., & Dickman, M. J. (2022). Characterization and sequence mapping of large RNA and mRNA therapeutics using mass spectrometry. *Analytical Chemistry, 94*(20), 7339–7349. https://doi.org/10.1021/acs.analchem.2c00765

Vlatkovic, I., Ludwig, J., Boros, G., Szabó, G. T., Reichert, J., Buff, M., Baiersdörfer, M., Reinholz, J., Mahiny, A. J., Şahin, U., Karikó, K., Vlatkovic, I., Ludwig, J., Boros, G., Szabó, G. T., Reichert, J., Buff, M., Baiersdörfer, M., Reinholz, J., … Karikó, K. (2022). Ribozyme assays to quantify the capping efficiency of in vitro-transcribed mRNA. *Pharmaceutics, 14*(2). https://doi.org/10.3390/pharmaceutics14020328

Wang, Y., Li, L., Kong, J., Yerabolu, R., Hullen, K., Zhao, K., Wen, E., Gunsch, M. J., Foley, D., & He, Y. (2025). DNAzyme approach for simultaneous mRNA cap and poly(A) tail length analysis: A one-step method to multiple quality attributes. *Journal of Pharmaceutical and Biomedical Analysis, 257*, 116695. https://doi.org/10.1016/j.jpba.2025.116695

Webb, A. L. J., Welbourne, E. N., Evans, C. A., & Dickman, M. J. (2025). Characterisation and analysis of mRNA critical quality attributes using liquid chromatography based methods. *Journal of Chromatography A, 1745*, 465724. https://doi.org/10.1016/j.chroma.2025.465724

Welbourne, E. N., Loveday, K. A., Nair, A., Nourafkan, E., Qu, J., Cook, K., Kis, Z., & Dickman, M. J. (2024). Anion exchange HPLC monitoring of mRNA in vitro transcription reactions to support mRNA manufacturing process development. *Frontiers in Molecular Biosciences, 11*. https://doi.org/10.3389/fmolb.2024.1250833

Wolf, E. J., Grünberg, S., Dai, N., Chen, T.-H., Roy, B., Yigit, E., & Corrêa, I. R. (2022). Human RNase 4 improves mRNA sequence characterization by LC-MS/MS. *Nucleic Acids Research, 50*(18), e106. https://doi.org/10.1093/nar/gkac632

World Health Organization. (2021, December 7). *Evaluation of the Quality, Safety and Efficacy of Messenger RNA Vaccines for the Prevention of Infectious Diseases: Regulatory Considerations.* https://www.who.int/publications/m/item/evaluation-of the quality safety and efficacy of messenger rna vaccines for the prevention of infectious-diseases-regulatory-considerations

Youssef, M., Hitti, C., Fulber, J. P. C., Kamen, A. A., Youssef, M., Hitti, C., Fulber, J. P. C., & Kamen, A. A. (2023). Enabling mRNA therapeutics: Current landscape and challenges in manufacturing. *Biomolecules, 13*(10). https://doi.org/10.3390/biom13101497

Żak, M. M., & Zangi, L. (2025). Clinical development of therapeutic mRNA applications. *Molecular Therapy, 33*(6), 2583–2609. https://doi.org/10.1016/j.ymthe.2025.03.034

Index

For Product Safety Concerns and Information please contact our EU representative GPSR@taylorandfrancis.com Taylor & Francis Verlag GmbH, Kaufingerstraße 24, 80331 München, Germany